EUROPEAN COASTAL ZONE

To

Dr. A. De Leeuw

Past President of the European Faculty of Land Use and Development

European Coastal Zone Management

Partnership approaches

Edited by

ROBERT W. DIXON-GOUGH
School of Surveying, University of East London

Routledge
Taylor & Francis Group

LONDON AND NEW YORK

First published 2001 Ashgate Publishing

Reissued 2018 by Routledge
2 Park Square, Milton Park, Abingdon, Oxon OX14 4RN
711 Third Avenue, New York, NY 10017, USA

Routledge is an imprint of the Taylor and Francis Group, an informa business

Publisher's Note
The publisher has gone to great lengths to ensure the quality of this reprint but points out that some imperfections in the original copies may be apparent.

Disclaimer
The publisher has made every effort to trace copyright holders and welcomes correspondence from those they have been unable to contact.

A Library of Congress record exists under LC control number : 00134081

ISBN 13: 978-1-138-63723-8 (hbk)
ISBN 13: 978-1-138-63732-0 (pbk)
ISBN 13: 978-1-315-20524-3 (ebk)

Contents

List of contributors

Professor Dr. W. H. Balekjian. Vice-Dean of the European Faculty of Land Use and Development, School of Law, Glasgow University.

Dr. R. K. Bullard. Land Reform Research Unit, School of Surveying, University of East London. *(Now at Anglia Polytechnic University).*

Dr. M. Déjeant-Pons. Directorate of Environmental and Local Authorities of the Council of Europe.

R.W. Dixon-Gough. Land Management Research Unit, School of Surveying, University of East London.

D.C. Doughty. Researcher, School of Surveying, Kingston University.

T. J. Goodhead. Department of Land and Construction Management, University of Portsmouth.

S. Goodman. University of Portsmouth.

A. Inder. County Planning Officer, Hampshire County Council.

S. Jaffry. University of Portsmouth.

Associate Professor J. Kioussopoulos. Technological Educational Institute, Athens.

Professor Dr. Dr. h.c. mult H. Lenk. Dean of the European Faculty of Land Use and Development, Institute of Philosophy, University of Karlsruhe.

Dipl.-Ing. U. Lenk. Project manager, University of Hanover.

Assistant Professor Dr. F. Narli. Institute of Marine Sciences and Management, Istanbul.

Professor W. Seabrooke. University of Portsmouth.

Professor Dr. P. Trappe. Department of Sociology, University of Basle.

J. Taussik. Department of Land and Construction Management, University of Portsmouth.

The European Faculty of Land Use and Development

Dr. A. De LEEUW

The European Faculty of Land Use and Development (Faculté Européenne des Sciences du Foncier; Europäische Fakultät für Bodenordnung) was founded in 1980 at Strasbourg. It has a mission to study and to develop interdisciplinary scientific methodology concerned specifically with the problems of land use and development across Europe, including such themes as urbanism, land management and the environment. In addition, its aim is to contribute towards the gaining of a greater degree of understanding of aspects of national legislation, across Europe, through the wide diversity of disciplines of the Faculty. This mission is accomplished by means of education and higher level research within the university sector throughout Europe.

The Faculty is a forum for discussion, comparison and exchanges of ideas and experiences, studies and research across those subjects within its broad scope. Within the framework of these activities, the Faculty co-operates with universities and research organisations both within Europe and internationally. The dissemination of this research is through the organisation of regular symposiums.

An Administration Council, for whom I have the honour to be its President, directs the Faculty: the official location being in Strasbourg (9, Place Kléber). It operates with the permission of the relevant French authorities and is a legal French association. The status of the association dates from the 12th December 1979 and it was registered in the "Registre des associations" of the "Tribunal d'Instance de Strasbourg" on the 17th September 1980. The Faculty is, at present, a group of 36 professors and readers, attached to 30 European universities. Each member is a recognised specialist in an aspect of land use and development.

To date, twenty-nine symposiums (each of duration of between 2 and 3 days) have been organised, usually twice a year. Each symposium normally includes a technical excursion related to the theme of the symposium. Three symposiums have been organised by myself in Strasbourg, seven in collaboration with the Council of Europe and nine in collaboration with the following universities: Granada, Cambridge, Lublin (Poland), the Royal Agricultural College (Cirencester), Vienna (twice at the Universität für Bodenkultur), Yildiz (Istanbul), Zurich (twice at the Eidenössische Technische Hochschule), Delft, Piraeus, Kingston (twice), Munich (Technische Universität), Bonn (in collaboration with the European Centre of the University of Waseda, Tokyo), Olsztyn (Poland), the University of East London, Sopron (Hungary), Portsmouth and Salzburg. Likewise, the study visits have been organised to Graz (Austria) at the Alfred Pikalo Institute.

The location and themes of the last five Symposiums have been:

No. 25 May 1997	University of Sopron (Székesfehévar) *"Land use Policy for Specific Applications"*
No. 26 September 1997	University of Portsmouth *"Coastal Zone Management. Partnership approaches to CZM"*
No. 27 May 1998	Eidenössische Technische Hochschule, Zurich *"Regional/Spatial Planning in a State of Flux"*
No. 28 September 1998	Kingston University *"Land Use Demand for Leisure and Recreation"*
No. 29 September 1999	Institut für Verfassungs- und Verwaltungsrecht, Universität Salzburg *"Land Law Systems in Urban Areas of Density and Alpine Regions"*

It is notable that the activities of the Faculty are carried out without any official financial support and only exists through the grace of some private sponsors.

The choice of the themes for the symposiums must address one or more of the following three criteria; a study of the topical subjects concerned with land use and development, to discuss the problems in an interdisciplinary manner, and to place the work in a European context. Our idea of Europe is not that of the European Union or even the Council of Europe, but one that stretches from the Atlantic to the Urals.

The working method of the Faculty is that a gathering, of up to a maximum of 20 to 25 participants, will present their papers - without interpretation - in one of the three working languages; English, French and German. The papers are discussed and, when possible, the reviewed and edited papers are published. To date, Peter Lang (Frankfurt am Main, Bern) has published 21 volumes (4 monographs and 17 symposiums). However, because of financial difficulties, the publication of the individual symposiums was suspended and this new formula of presentation and publication is being adopted through Ashgate Publishing.

The Faculty's relationship with other Universities outside Europe should also be pointed out, notably Waseda University of Tokyo and the University of Buenos Aires and Cordoba in Argentina. Thus, the Faculty has also been instrumental in playing a part in the scientific exchanges between Europe and those countries.

This present volume contains the papers presented at the 26th Symposium organised in co-operation with the University of Portsmouth, the theme being "Coastal Zone Management: Partnership Approaches to CZM".

Strasbourg, June 1998

Dr. A. De Leeuw
Président du Consiel d'Administration de la Faculté Européenne

1 Partnerships in European coastal zone management

R.W. DIXON-GOUGH

No man is an Island, entire of it self; every man is a piece of the Continent, a part of the main; if a clod be washed away by the sea, Europe is the less, as well as if a promontory were, as well as if a manor of thy friends or of thine own were; any man's death diminishes me, because I am involved in mankind; And therefore never send to know for whom the bell tolls; it tolls for thee.

John Donne (1572-1631)
Devotions upon Emergent Occasions (1624) 'Meditation XVII'

Introduction

Although the words of John Donne were written almost 400 years ago, and as a metaphysical poet probably related to something quite different, they nevertheless manage to synthesise many of the concepts related to the interrelated and linked phenomena and processes around the coastline of Europe. The coastline of Europe has been described as one of its last wildernesses, a finite resource and a boundary between the land and sea that is of international importance (Smart, 1992). The term 'Coastal Zone' is normally taken as referring to the interrelated landward and seaward components of the coast and may be defined as

> a zone based on environmental factors first, then refined in the light of administrative and practical considerations (Marine Conservation Society).

It is characterised by a complex association of resources, pressures and opportunities. Within this zone, there are many competing human interests each having a different perspective concerning the nature of any opportunity or problem. Also linked to those human interests are physical and environmental processes.

Many authors, for example French (1997) categorise the physical components and processes of the coast. These interrelated and linked phenomena and processes often defy categorisation since each component, whilst certainly linked to others, is nevertheless a unique entity with its own set of criteria, problems and solutions. The greatest difficulty lies in attempting to develop a set of procedures that will fit within a national or international framework that will permit those entities and their population to feel that they can exercise some degree of control over their decisions. Smart *op cit* gives examples that defy many conventions relating to the categorisation of coastal forms. In the case of the 'hard coastline' he gives the accepted definition of hard, vertical cliffs that are often located in highland Britain and are characterised by a robust landscape and wildlife resource that is often of international importance. This may be contrasted with another form of hard coast, that of an urban and industrial region that is both an economic and social resource, which might in turn give rise to recreational pressures upon the adjacent coastal region. This contrasts considerably with the concept of the 'soft' coastline, which is very sensitive to misuse and anthropogenic pressures. Much of the soft coastline of Europe consists of estuaries that have internationally recognised wildlife value and these are very vulnerable to the dynamics of land-induced changes.

In addition to general category of anthropogenically-induced changes, the coastline is also being continually subjected to the process of natural change. If we accept that the land as being a sedimentary budget for the sea, we must also accept the natural processes of change, erosion and deposition. Much of the coastline of Europe is protected from the sea and this poses an environmental, scientific and visual threat. Certainly in England and Wales, much of the low-lying coastal areas have been reclaimed and protected, some of which date back to Roman times. Historically, there have always been competing demands for space on the coast, particularly in the fertile, flat areas of the 'soft' coastlines. Of the estuarine regions of the UK, it is estimated that as much as 25% of intertidal land has been lost since Roman times (Davidson *et al.*, 1992). More recently, it has been estimated that 64% of grazing marsh in the greater Thames area has disappeared since the 1930s (Clarke *et al.*, 1991). Furthermore vast areas have been reclaimed for industrial purposes and these area, because of their increase in 'value' now require protecting, particularly in the light of the predicted rises in sea level and increased frequency of storms. While certain problems within the coastal zone have increased, largely as the result of human exploitation, conflicts between the human and natural environments have become more apparent as human dependence on Europe's coastal resources has increased.

The coast is also increasingly being used as an industrial resource. There are currently plans to extend and build container ports in the Test estuary at Southampton, and at Felixstowe. Both of these proposed sites are on intertidal areas that are internationally recognised wildlife areas. Then there are the ecologically and visually damaging effects of aquaculture in many of the estuaries and sea lochs and fjords of Scotland and Norway. Other effects may be less localised and the effects of pollution from industrial sources, the discharge of sewage and the oil and chemical spills may all be widespread both in terms of the visual effect, environmental problems and upon local economics. Gilbert (1992) has identified a number of different sources of pollution that effect the coastline of Kent. Some of these have been generated locally, such as PCBs from the paper mills at Thamesside and the Medway area, the acidification of coastal waters as the result of sulphur dioxide emissions from Thamesside, and discharge of untreated sewage into coastal waters. Other sources of pollution are far more remote and outside the jurisdiction of Kent or even the UK. In the waters off coastline of Kent, during the early 1990s, significant amounts of lindane was being detected. The origin of this chemical, extensively used on the collective farms of East Germany and Poland from the 1950s, was from the Waddensea and from there into the North Sea. It had been originally deposited in the Waddensea as the result of run-of from the river Elbe. It is for such consequences as this that some form of European-wide partnership must be agreed for coastal zone management throughout the 'Greater Europe'. Significantly, the county of Kent is now located within a Euroregion, which comprises parts of Nord Pas de Calais and three Belgian regions, all of which are geographically closer to the shoreline of Kent than, say, the coastline of certain parts of Essex.

What is indisputable is that there is a 'state of conflict' within the coastal zone, which has now reached the political arena and may have, in additional to environmental implications, severe social and economic consequences (Sidaway, 1995). In case studies that have been conducted throughout Europe there seems to be evidence of a gap between frequent widespread support for coastal conservation but not when the objectives of a coastal conservation programme is applied to specific and localised conservation measures. It is therefore very important to develop a local strategy plan, which involves the active participation of local interest groups in both its evolution, development and implementation. One such initiative is the Morecambe Bay Strategy Plan (Dixon-Gough, 1998).

Towards a coastal zone management system

The need for a specific management system to address the unique problems of the coastal zone arises from the need to integrate the management processes of the land and sea. A successful coastal zone management system must be based on a decision making process capable of integrating all relevant issues and interests (Gubbay, 1990). In most cases the pressure is, in anthropogenic terms from the land to the sea, whilst in natural and ecological terms from the sea to the land. Furthermore, it has been estimated that 50% of the population of the industrialised world lives within 1 km of the coast and that this is likely to increase at the rate of 1% per annum over the next decade (Goldberg, 1994). Within greater Europe, many of the largest cities are located directly on the coast, while many others are very close to the coast, often on rivers or within estuaries. These concentrations of population can impact both directly and indirectly upon the coastal zone. In a direct manner, they have to be supported by an industrial, commercial, agricultural and service infrastructure but indirectly, the may influence the coastal zone through tourism and by the increasing needs for quality residential developments, in prime coastal locations.

Interlinked with those concentrations of populations are twin threats of increased recreation time and tourism, which has led to widespread and increased use of the coasts and estuaries throughout Europe (Davidson & Rothwell, 1993). This can only add to the pressure currently being experienced within the coastal zone of Europe and will give rise to many concerns, notably upon the effects of this population increase upon the limited coastal resource. This resource is both fragile and finite and will include the issues of coastal erosion as well as the needs for coastal defence to protect land and property. This will impact upon habitats, create additional problems concerning all types of pollution, which will inevitably give rise to a reduction, both in absolute and relative terms, in bio-diversity (Healy, 1995).

As the result of such pressures the need for an integrated management structure is emphasised by the increased intensity of use of both the land and sea in the immediate vicinity of the shore (Smith, 1989). The greatest criticism relating to the concept of coastal zone management is that in many European countries there is no legislative framework. In the case of Great

4

Britain, there is not even a map series that spans the coastal zone in a single geometric framework: the Ordnance Survey map to the low water marks and the Hydrographic Office to the high water marks but neither use a common geometrical framework and neither do they share common data.

Future coastal zone management throughout Europe will necessitate an integration of existing knowledge and research, together with the harmonisation of regulations, legislation and human interest, in order to provide a cohesive and well-structured framework. A starting point for this process is to be able to document the nature and pattern of coastal change and to relate this to current physical, environmental and anthropogenic processes. The 1992 United Nations 'Earth Summit' conference at Rio de Janeiro, expressed concerns and initiated an international framework for environmental protection and sustainable use of the environment. As a result of an increased acceptance of the concepts of 'sustainability', largely as a direct result of this conference, and of 'sustainable coasts', many involved in the activities relating to coastal management have recognised that the coastal zone is not the preserve of any one activity (Doody, 1995).

Whilst coastal conservation is not the same as coastal zone management, the legislation for coastal conservation can provide a very useful framework for such a purpose. Nordberg (1995) provides an interesting comparison of the nature of coastal conservation and the associated legislation and regulation in selected European States. He concludes that there is little chance of harmonising the existing legislation on a European level since both natural and political circumstances prevent this. He also comments that since coastal conservation is a pan European issue, other European structures should be investigated as offering a potential solution. The challenges to coastal managers during the early years of the new century will be to develop and adopt new frameworks and methodologies that will allow the human interests to be harmonised with those of the environment and nature (Tooley & Shennan, 1987).

One piece of legislation that might provide a platform for the development of a partnership approach to coastal zone management across Europe is the European Union Habitats Directive (European Council Directive 92/43/EEC) on the conservation of flora and fauna and their habitats. The formal process of identifying and designating Special Areas for Conservation (SACs) and Special Protection Areas (SPAs) within the Natura 2000 network will provide an opportunity to add rigour to the planning and management of conservation practices in most European countries. Under the Habitats Directive, each member state of the European

Union is required to establish, within its own legislative framework, a means of conserving the flora and fauna. This requirement directly addresses the issues involved in reconciling the conflicting and cultural conditions prevalent in each member state, within a common European framework and, as such, can only benefit the development of a pan European coastal zone management framework.

It is through the recognition of legislation such as the Habitats Directive and conferences such as the 'Earth Summit', that it has become accepted that the current level of exploitation of the coastal zone cannot be sustained. This has, in turn, led to the development of a more comprehensive approach to its management. One such approach is through Integrated Coastal Zone Management (ICZM), which has developed since the mid 1970s as a tool for managing the use and exploitation of the coastal zone in many parts of the world. This tool was extensively promoted by the United Nations through its Environment Programme, the IUCN and also by the Council of Europe (Amselek *et al.*, 1994). The implementation of any legislation is effectively 'bottom up' and unless a significant level of co-operation can be established between the legislative measures and the local people, upon whom the success of the plan will succeed or fail, the programme will be unlikely to be a great success. The keystone to this process is the development of partnerships that will provide a direct link between the legislators, implementers and the users.

Partnership approaches

This book is a presentation of edited papers that fall into five main groupings. Firstly, Déjeant-Pons, Trappe and Dixon-Gough address the general situation, philosophy and policies of coastal zone management within a European context. Chapters by Lenk & Lenk, Goodman, *et al.* and Goodhead emphasising the inter-disciplinary nature of coastal zone management follow this. In the next grouping of chapters, Taussik examines more specifically, the legal and regulatory framework for planning within England and Wales whilst Inder gives a case study of the partnerships involved in the planning and management of the Solent region of southern England. More specific case studies are presented by Bullard and Doughty who respectively regeneration of saltmarshes and the declining English coastal resort, whilst Balekjian describes the need for an integrated coastal zone management partnership within the Baltic Sea. The final two papers

6

identify the lack of integrated structures for coastal zone management in Greece and Turkey, respectively.

One of the driving elements behind coastal zone management is the need to either establish or re-establish the balance between nature and the demands of an increasingly urbanised and recreation-driven society. One of the organisations driving the quest towards European-wide partnerships in the common heritage of all European states is the Council of Europe. This founded in 1949 following the adoption of the Council of Europe Statute in London. Déjeant-Pons outlines the evolution of this organisation and describes its influence upon pan-European awareness of the environment of Europe, in general, but more specifically its influence upon the development of partnerships in coastal zone management. A number of instruments and conventions adopted by the Council of Europe are described and it is significant to note that in 1973, Resolution (73) 29 concerning the protection of coastal areas, was adopted by the Committee of Ministers. This took into consideration the serious biological degradation of Europe's coastline and noted the indiscriminate siting of buildings, infrastructure and tourist facilities, particularly in the Mediterranean basin. One of the principal effects that the Council of Europe has had, more recently, upon the coastal zone has been through the Pan-European Biological and Landscape Diversity Strategy. This sets out to achieve the objectives of conserving, enhancing and restoring key ecosystems and landscapes, whilst promoting their sustainable management at local, national and regional levels. Part of this strategy specifically addresses coastal and marine ecosystems, and the development of an integrated coastal zone management approach to the utilisation of the land and sea, based upon the considerations of conservation.

In the second chapter, Trappe takes a philosophical look at the complementary concepts of de-nationalisation and self-help with respect to the development of partnerships in coastal zone management. Whilst this might initial appear to be at odds with the concepts outlined by Déjeant-Pons, it must be remembered that all planning and conservation process operate, as opposed to being legislated for, at a local level. This is in fact emphasised by the Council of Europe's Pan-European Biological and Landscape Diversity Strategy, which emphasises that the thrust of sustainable management is at local, national and regional levels.

Dixon-Gough identifies the concepts of coastal zone management and the relationship between management frameworks and the needs of local communities. One of the inescapable facts is that no matter how local is the strategy and implementation of a coastal zone management framework, the political dimension will be the major controlling factor. One of the most influential events in recent years, which has helped place coastal management high on the political agenda, was the UN Conference on the Environment and Development (the Earth Summit) held in Rio de Janeiro in June 1992. This conference reconciled of the twin goals of economic growth and environmental protection under the general political focus of 'Sustainable Use'. In this context, Agenda 21 identified the need for a comprehensive programme of action to achieve a sustainable pattern of development over the next 100 years and has formed the basis for many regional, national and local initiatives. Also at a global level is the United Nations Environmental Programme (UNEP) has become a leading force in the development of international environmental programmes. One such programme addresses the management of the marine environment through its Regional Seas Programme. A regional approach to the control of regional pollution and the management of marine resources was endorsed and this was initiated as the Regional Seas Programme. One such initiative that has developed from this programme is the Mediterranean Action Plan (MedPlan) has brought together all countries within the Mediterranean basin for a common purpose and goal and has achieved a high level of regional co-operation.

Since the beginning of 1992 the European Union has been working on this but no European legislation seems to be in sight in the near future.

As a European level, no instrument *per se* exists regarding coastal conservation and therefore the potential of other European structures may be investigated. For example, the Helsinki Convention (Convention for the Protection of the Marine Environment of the Baltic Sea Area) now provides provisions concerning nature conservation and bio-diversity (Article 15). However, for a pan-European framework, more precise rules than those in the Helsinki Convention would be needed. In principle, the Bern Convention is suitable for this, although until now it has been concerned mainly with the protection of individual species and coastal conservation and landscape protection are not considered. More recently, the Pan-European Biological and Landscape Diversity Strategy, as described in detail by Déjeant-Pons, may offer a more refined solution.

Coastal Zone Management (CZM) has been on both national and European Union (EU) political agendas now for some 25 years and as early as 1970, changes to the way in which the coastline of England and Wales was managed were recommended by the Countryside Commission. In response to these concerns, the Council of Environment Ministers adopted a resolution calling for the production of an EU strategy for integrated coastal zone management (ICZM) framework and its incorporation into the 5th Environmental Action Programme. Unfortunately, progress was slow and in 1994, the Council of the European Union renewed its invitation for the Commission to develop an ICZM strategy for the entire coastline of Member States. In 1995, the Commission published a Communication on CZM, which proposed the establishment of a demonstration programme. The results of the demonstration programme were finally published in 1999 and some of the more significant aspects of this programme will be outlined below.

The interdisciplinary nature of coastal zone management is emphasised by Lenk & Lenk who throw fresh light on the concept of partnerships, particularly in the disciplines of planning and management. The role of experts and practitioners from very different fields is not exclusively restricted to coastal zone management but in many environmental and ecological roles. Lenk & Lenk draw a number of similarities between the management of the coastal zone and floodplain area development, both of which are being subjected to increasing demands and both to a combination of environmental, social and economic pressures. This chapter discusses the role of geographical information technology as a means of providing timely data and information in conducting, structuring and controlling coastal zone management developments and networks. Any geographical information system (GIS) is an interdisciplinary tool capable as acting as a link between the disciplines involved in coastal and floodplain management but also in relating such diverse aspects as research, theory and application. However, the human aspects of coastal and floodplain management cannot be ignored and these have to be used to supplement and be integrated within any modelling process to expand the social, environmental and economic elements.

Those social, environmental and economic elements are increasingly being used in an attempt to calculate the total economic value of natural resources, together with an economic assessment of policy options, of the coastal zone. Goodman *et al*, suggest that it is only by making such an analysis that 'worth' of an entity and the justification of a management strategy and framework can be assessed. Both valuers and economists are reasonably confident of their ability to set a realistic price, which users would realistically be prepared to pay to exercise their right to use natural resources. However, the cultural tradition of depicting the natural environment, particularly in the case of new residential areas and tourism, has often emphasised certain qualities such as the freshness of the air, the clarity of water, the richness and variety of habitat. The authors consider that in order to develop a rigorous basis for estimating the total economic value of a wide range of coastal resources, a study must be made to explore the use of a conservationist valuation framework to assess preferences for stereotypical features of the British coast. This framework should identify and provide a relatively objective basis for the evaluation of the physical characteristics that give the natural resources of the coast a conservation value. Similarly, it should identify criteria that are widely accepted as being significant in determining that conservation value.

The hypothesis suggested by Goodman *et al* was to establish a direct relationship between non-use and conservation values that could indicate that conservation quality levels could be used as a basis in the attribution of non-use values to coastal resources. Conservation values could be estimated for a broad range of site characteristics, including those, which are not considered to be of particularly outstanding quality i.e. ubiquitous rather than unique. However, this approach would almost certainly overestimate the benefits derived by the British public from coastal conservation quality and might have a profound influence upon the development of any coastal zone management framework.

Goodhead who notes that conflicts in coastal areas and perhaps to a lesser extent at sea, have become more severe over recent years develops this approach further. Factors such as coastal protection, large scale clearing, urban development and aquaculture have led to valuation problems within the coastal zone. Thus, the management of the coasts of Europe is creating many problems and, indeed it could be argued, that there has been a policy failure in that there has been little policy. However, awareness of the issues in coastal management is so new that it may be unfair to be too critical of policy although there does seem to be a lack of political commitment. Policy has almost developed on an *ad hoc* basis and, as a result, there are often conflicts between the objectives of individual agencies, rather than the co-operation and partnership agreements so ideally required. The formation of a policy towards coastal zone management is often difficult since the data available are often either non-available, inconsistent or very basic. In many instances, the methods of data collection are not comparable within a region, let alone the whole of Europe, due to the overlapping responsibilities of the agencies, which can mean that the mechanisms of intervention are often unsuitable and frequently in conflict with one another. However, the concept of 'ecologically sustainable development', developed mainly from UN Conferences, together with

Natura 200 networks or the European Union and the Council of Europe's Pan-European Biological and Landscape Diversity Strategy has started to provide a broad framework for developing coastal and marine management systems. Thus sustainable developments of our resources have become very topical.

Goodhead proposes the development of network and databases of those professionals and interest groups active within the coastal zone management sector. For example, within the UK, a number of marine and coastal networking forums exist, such as CoastNet, the National Coast and Estuaries Advisory Group and the European Union for Coastal Conservation. The most notable of these is probably CoastNet, a coastal heritage network, which has produced a UK Coastal Management Directory (CoastNet, 1997) to aid communications, networking and exchanges between those involved with the UK's coastal and marine environment. Within these networking organisations, there appears to be a clear role for a chartered or licensed surveyor both in Marine Resource Management and property professional, generating business in the coastal and marine zones. However, due to the complex nature of the policies within the coastal zone, success in this area may be a function of the ability to liase and network with other professionals to form multi-disciplinary teams and, as Goodhead notes, effective networking and marketing is vital.

Although networking promises a partial solution to the problems of coastal zone management, Taussik suggests that a further solution might be developed through the progress achieved in implementing policies of integrated coastal zone management. This could be based upon guidance prepared by international agencies in order to resolve conflicts of interest in the coastal zone. However, coastal zone management operates in various contexts throughout Europe, including variations in geographical scale, organisational scope, level of horizontal and/or vertical integration, the time horizon and the level of community involvement. One of the greatest problems in developing any international partnership across European countries is that there is a lack of statutory frameworks within the individual states. For example, within the UK it has been accepted at governmental level that there is a need for coastal zone management. However, it was not considered feasible to consolidate the legislation affecting the coastal zone or to establish a national coastal zone unit to adopt a national overview of coastal zone policy. Furthermore, it was not believed that there was a widespread duplication of responsibilities or poor co-ordination in coastal management. The advantages of integrated approaches are increasingly recognised but initiatives in the UK can only be undertaken through non-statutory, voluntary means. Some of the partnerships and

11

initiatives currently being undertaken are outlined by Taussik who also describes the UK's planning system and more recent acceptance of coastal zone management operated through voluntary means.

Although Taussik looked at the general problems of developing a countrywide or internationally based partnership in coastal zone management, Inder concentrates upon the level and diversity of partnerships necessary to implement a relatively localised scheme. As for any other part of the coastal zone within the UK, the basic organisational structure for planning and management in the Solent, in southern England, is highly fragmented. The types of statutory and non-statutory authorities and agencies that have a role to play, each with its own powers, objectives and priorities, and area of jurisdiction are clearly identifiable and even within this grouping of authorities there are significant variations. For example, in the Solent, there are eight harbour authorities alone, all with their own powers and responsibilities, whilst the main shipping channel in the western Solent is outside the jurisdiction of any harbour authority. The structures identified by Inder clearly indicate the levels of complexity in developing a coastal zone management framework within the UK, let alone Europe. Similarly, the planning procedures within the region are based largely on the statutory Town and Country Planning system, which is restricted to land use and development and is unable to tackle management issues. An additional limitation is that from the coastal point of view, jurisdiction normally ends at the low water mark, although in the Solent there are significant exceptions. The pressures on the area are increasing, too and this phenomenon has been experienced, almost without exception throughout Europe. Since 1951 the population of the Solent area has increased by 34% and the boat numbers have increased by almost 45%, whilst although the number a merchant shipping movements have decreased, the overall trade handled by the ports of Portsmouth and Southampton has increased significantly in recent years. Inder comments that these pressures have increasingly exposed the inadequacies of a fragmented system of administration and have led to calls for a more integrated system of planning and management, and a more long-term, strategic approach to localised coastal zone management.

Bullard continues the in the theme of identifying local problems relating to the coast by addressing the issues of coastal defence. In Europe responsibility for the coastline is largely the responsibility of the central government with this authority being delegated to water authorities and other statutory bodies. In the United Kingdom, the Crown Estates own about 55% of the foreshore and almost all the seabed out as far as the territorial limit. Another significant owner of the coastline of the UK is the National Trust, a charitable body, with a responsibility for the purchase and subsequent management of land of major importance to the country's heritage. Responsibility for the coastline is further complicated by the jurisdiction of various organisations that overlap considerably. One of the greatest problems associated with the coast is its defence and who will take the responsibility for that defence. Since the geology of a coast will determine whether it is a soft or hard coastline, any attempt to protect a soft coastline will serve to make that coastline harder and more rigid, often to the detriment to adjacent stretches to the coast. Bullard takes the specific example of a soft coastline and examines a particular case study in which the Environment Agency is practising a form of measured retreat. In this case, reclaimed saltmarsh is being allowed to return to a natural state by the removal of the coastal defences, thereby forming part of a 'new', soft coastal defence system. This can only be achieved through a full partnership between landowners and the agency responsible for coastal defence and may offer a possible means of providing a more natural and effective system of coastal defence in certain areas.

Doughty who documents the phenomenon of the British coastal resort, in this case gives a further example of localised coastal issues, that of Lytham St. Annes in north west England. Such resorts grew in popularity during the nineteenth century but, largely as the result of changing fashions and attitudes, have become neglected as the greater affluence of holiday-makers seek out the assured sun of the Mediterranean resorts. This has, in turn, generated other problems of coastal zone management within the Mediterranean basin. What has been experienced in Lytham St. Annes has also been experienced in many other holiday resorts in Britain and is largely due to a lack of investment in infrastructure. Some resorts have continued to be popular but others will gradually decay and eventually create a different type of coastal problem. Indeed, some may well go the way of the reclaimed saltmarshes described by Bullard, since as their economic value declines, the arguments for expensive coastal protection to defend them will diminish.

In contrast to the localised problems described by Bullard and Doughty, Balekjian discusses the wider contexts of a region, that of the Baltic Sea, which is bounded by a number of states. The Baltic is essentially an inland sea with a mixture of almost dilute salt and almost sweet water contents. Furthermore, it could be also described as a complex fjord, firth or gulf with complex subdivisions, separated by many low seabed thresholds. It is a relatively young sea that has been a navigational link between its ethnically and politically different communities and is now becoming, for the purpose of coastal management, an integrated area within the larger framework of the European Union and European co-operation in general, as discussed also by Déjeant-Pons and Dixon-Gough.

The most important challenge the region has to undertake, in both an industrial and economic context, is that coastal management in the Baltic area has to assume macro-dimensions. The factors influencing the Baltic Sea extend far beyond the immediate geographical dimensions of coastal regions. They comprise an area, which extends up to 400 kilometres southward into Central Europe and equally far eastward unto Russia, beyond the three small Baltic republics. This is due to major rivers like the Vistula (Poland) and Oder (Germany and Poland) serving as drainage channels for industrial and other emissions. The river Vistula has been carrying industrial effluents and polluted waters from some 400 cities, villages and industrial areas. The development of any coastal zone management framework for this region will rely heavily upon trust and partnerships between both the states bordering the coast of the Baltic Sea and those states through which rivers pass, which drain into the sea.

The final two chapter deals with problems that are specific to the development of coastal zone management frameworks in their countries. Kioussopoulos outlines the current situation in Greece in which he identifies the need for multi-sectoral approach to coasts in order to give the necessary umbrella for wisdom and effective actions. However, he accepts that the very long Hellenic coastline, one of the longest in Europe, needs a management framework in order to regulate the conflicting land uses along it since the previous general efforts in the field of coastal policy have no visible results. In this chapter, Kioussopoulos outlines the structure of a database aiming at the monitoring and the estimation of land use changes in the Hellenic coastal area. To date, the development has primarily concentrated upon the development of an appropriate framework. This would allow most of the represented variables to be identified together with the most useful data sources. As land uses are the outcome of competition of many factors, the framework and the database will provide a starting point for the organisation of the needed interdisciplinary contacts and other linkages. It is anticipated that this database will not only form an inventory of the present coastal situation, but also to offer substantial and continuous assistance to the decision-makers concerning the coastal zone of Greece.

Narli describes the current problems in Turkey where planning, application and monitoring of the plans and furnishing of the public services within the coastal zone have not been handled in an integrated management strategy. This is not a situation that is confined to Turkey but may be found in most European countries. Nevertheless, Nalri makes a very thorough analysis of the problems encountered in Turkey with respect to the management of developments and the environment within the coastal zone. In Turkey there is no single national organisation that is capable of co-ordinating the activities of the institutions responsible for the various activities within the coastal zone. Indeed, this problem is encountered throughout Turkey and there is no structure for the management of development politics at a national scale, whilst addressing the demands, identifying targets and programming priority actions. Thus, there are no regional and local organisations to which the authorities responsible of coastal zone management have been defined. Therefore, the problems and needs of these areas, determining the priority regions and the settlements, preparing the priority action plans, regional development and management programmes have not been addressed in a structured manner. Thus, within Turkey, there are many different organisations responsible for the various steps of the management process. These include the Governmental Planning Authority and some ministries, together with their general directorates. Also, general directorates associated with public services may participate the management without any organised collaboration among them and without having a well-defined national or regional management policy and programme. Whilst this situation is not untypical, it is, nevertheless an indication of the serious nature of coastal zone management throughout Europe. Furthermore, it is an indication of the need to develop a defining framework that addresses the issues of coastal zone management across Europe, whilst permitting the individual states to develop their own institutional arrangements in a way that satisfies their cultural and specific geographical requirements.

References

Amselek, P., Cohen, J. & Prieur, P., 1994. *Legislative Measures Taken or to be Taken by the Member States of the Council of Europe for the Protection of the Coastline,* European Information Centre for Nature Conservation, Council of Europe, Strasbourg.

15

Clarke, M., Bayes, K. & Durdin, C., (eds.) 1991. Must the Greater Thames be sacrificed, *Conservation Review*, 5, 88-93.

Davidson, N.C., Laffoley, D.d'A., Doody, J.P., Way, L.S., Gordon, J., Key, R., Drake, C.M., Pienkowski, M.W., Mitchell, R. & Duff, K.L., 1992. *Nature Conservation and Estuaries in Great Britain.* Joint Nature Conservation Committee, Peterborough, England.

Davidson, N.C. & Rothwell, P., 1993. *Disturbance to Waterfowl on Estuaries,* Wader Study Group, Bulletin 68, Special Issue.

Dixon-Gough, R.W., 1998. An analysis of the interaction between tourism and nature around Morecambe Bay, north-west England, 26th International Symposium of the European Faculty of Land Use and Development, *Land Use Demand for Leisure and Recreation,* Kingston University, London.

Doody, J.P., 1995. Information and coastal zone management. In: Healy, M.J. & Doody, J.P., (eds.). *Directions in European Coastal Management,* 399-414, Samara Publishing Limited, Cardigan, Wales.

French, P.W., 1997. *Coastal and Estuarine Management,* Routledge Environmental Management Series, Routledge, London.

Gilbert, C., 1992. Case study 3: the Kent coast strategy. In: Hampshire County Council, Proceedings of a Conference on *Coastal Planning and Management,* 20-22, Winchester, U.K.

Goldberg, E.D., 1994. *Coastal Zone Space: Prelude to Conflict?* UNESCO, Paris.

Gubbay, S., 1990. *A Future for the Coast? Proposals for a UK Coastal Zone Management Plan.* Marine Conservation Society/WWF, London.

Healy, M.J., 1995. European coastal management: an introduction. In: Healy, M.J. & Doody, J.P., (eds.). *Directions in European Coastal Management,* 1-6, Samara Publishing Limited, Cardigan, Wales.

Morecambe Bay Partnerships, 1996. *Morecambe Bay Strategy,* Morecambe Bay Partnerships, Grange-over-Sands.

National Rivers Authority, 1991. *The Future of Shoreline Management,* NRA Anglian Regional Conference Paper, Sir William Halcrow and Partners, Swindon.

Nordberg, L., 1995. Coastal conservation in selected European States. In: Healy, M.J. & Doody, J.P., (eds.). *Directions in European Coastal Management,* 47-50, Samara Publishing Limited, Cardigan, Wales.

Organisation for Economic Co-operation and Development, 1993. *Coastal Zone Management: Integrated Policies,* OECD, Paris.

O'Riordan, T. & Ward, R., 1997. Building trust in shoreline management, *Land Use Policy,* 14, 257-278.

Pethick, J., 1996. The sustainable use of coasts: monitoring, modelling and management. In: Jones, P.S., Healy, M.G. & Williams, A.T., (eds.), *Studies in European Coastal Management,* Samara Publishing Ltd., Cardigan.

Sidaway, R., 1995. Recreation and tourism on the coast: managing impacts and resolving conflicts. In: Healy, M.J. & Doody, J.P., (eds.). *Directions in European Coastal Management*, 71-78, Samara Publishing Limited, Cardigan, Wales.

Smart, G., 1992. Coastal planning and management: overview and up-date. In: Hampshire County Council, Proceedings of a Conference on *Coastal Planning and Management*, 5-12, Winchester, U.K.

Smith, H., 1989. Coastal zone management and planning: making it work, Proceedings of the Symposium of *Planning and Management of the Coastal Heritage*, 45-49, Southport.

Tooley, M.J. & Shennan, I., 1987. *Sea Level Changes*. Blackwell, Oxford, England.

2 Council of Europe activities concerning the protection of coastal zones

M. DÉJEANT-PONS

Introduction

> International law, as reflected in the provisions of the United Nations
> Convention on the Law of the Sea [.....] sets forth rights and obligations
> of States and provides the international basis upon which to pursue the
> protection and sustainable development of the marine and coastal
> environment and its resources. This requires new approaches to marine and
> coastal area management and development, at the national, subregional,
> regional and global levels, approaches that are integrated and are
> precautionary and anticipatory in ambit [...]
>
> (Agenda 21, adopted in Rio de Janeiro on 14th June 1992 by the United
> Nations Conference on Environment and Development, chapter 17,
> "protection of the oceans, all kinds of seas, including enclosed and semi-
> enclosed seas, and coastal areas and the protection, rational use and
> development of their living resources", para. 17.1)

The Council of Europe Statute, adopted in London in 1949, states that the
organisation's aim is to achieve a greater unity between its members for the
purpose of safeguarding and realising the ideals and principles which are their
common heritage and facilitating their social and economic progress. It is
provided that this aim will be pursued by discussion of questions of common
concern and by agreements and common action in economic, social, cultural,
scientific, legal and administrative matters.

The conservation of the European natural and cultural environment, the
common heritage of Europe's nations, very quickly became one of the main
problems of society, which Europe would have to face. This prompted the
Council of Europe to attach special importance to the matter and to develop
projects on the elements (seawater, fresh water, air and soil), land (natural, rural,

mountain, lakeland, river and coastal areas), animals and plants (wild, domestic, cultivated and genetically- modified species) and cultural assets (particularly architectural and archaeological assets). As it was essential in this context to take account of human activity and its environmental impact, the question of tourism and its connection with the environment was brought increasingly to the fore, and subsequently became the subject of specific activities.

With its current membership of 40 states and expected to take in other central and east European states in the near future, the Council of Europe is now trying to tackle the phenomenon of spatial management and its environmental impact at the level of Greater Europe. The pressures on central and east European countries are taken into consideration. The organisation's activities with Mediterranean states also give it a specifically Mediterranean view of things. The activities conducted within the Mediterranean Action Plan are also taken fully into consideration.

Activities consist of awareness-raising and information campaigns, political and administrative measures and obligations imposed by legal texts (conventions, charters, recommendations, resolutions and so on). These activities are conducted at different levels: parliamentary (Parliamentary Assembly), local and regional (Standing Conference of Local and Regional Authorities of Europe (CLRAE), which on 3 June 1994 became the Congress of Local and Regional Authorities of Europe) and intergovernmental (Committee of Ministers). Conferences of specialised ministers meet to define policy guidelines in their particular fields of competence.

1995 was declared European Nature Conservation Year (ENCY). More than forty countries took part in this pan-European campaign which aimed to promote nature conservation not only within but also outside protected areas, in other words, where social and economic activities are conducted. The closing colloquy, on 2 April 1996 at the Palais de l'Europe, looked especially at the future of environmental protection in the Mediterranean Basin.

The Council of Europe's action is part of a pan-European process of regard for the environment at ministerial level, 'An Environment for Europe', launched in 1991 at Dobris, continued in 1993 in Lucerne and in 1995 in Sofia, and to be pursued in 1998 in Aarhus, Denmark.

The Pan-European Biological and Landscape Diversity Strategy, adopted in Sofia on 25 October 1995 by the Ministerial Conference on the Environment, and implemented by the Council of Europe together with the United Nations Environment Programme (UNEP), has four aims:

- to substantially reduce and, if possible, to eliminate current threats to

Europe's biological and landscape diversity;
- to increase the resilience of Europe's biological and landscape diversity;
- to strengthen the ecological coherence of Europe as a whole; and
- to raise considerably the degree of public awareness and involvement in the conservation of the various aspects of biological and landscape diversity.

Activities conducted within the Pan-European Biological and Landscape Diversity Strategy

Broad outline of the Pan-European Biological and Landscape Diversity Strategy

The Pan-European Biological and Landscape Diversity Strategy aims to stop and reverse the degradation of biological and landscape diversity assets in Europe. 54 Member States of the United Nations Economic Commission for Europe (UN/ECE) are taking part in its implementation. These consist of; Albania, Andorra, Armenia, Austria, Azerbaijan, Belarus, Belgium, Bosnia-Herzegovina, Bulgaria, Canada, Croatia, Cyprus, Czech Republic, Denmark, Estonia, Finland, France, Georgia, Germany, Greece, Hungary, Iceland, Ireland, Israel, Italy, Kazakhstan, Kirghizistan, Latvia, Liechtenstein, Lithuania, Luxembourg, Malta, Monaco, Netherlands, Norway, Poland, Portugal, Moldova, Romania, Russia, San Marino, Slovakia, Slovenia, Spain, Sweden, Switzerland, Tajikistan, 'the former Yugoslav Republic of Macedonia', Turkey, Turkmenistan, Ukraine, United Kingdom, United States and Uzbekistan. The Strategy aims to promote concerted European action to protect the genetic diversity of wild and domesticated species through habitat measures. In this sense, it is a European response to promote the application of the Convention on Biological Diversity. The Strategy introduces a co-ordinating and unifying framework for strengthening and building on existing initiatives and programmes. It does not aim to introduce new legislation or programmes, but to fill gaps where initiatives are not implemented to their full potential or fail to achieve desired objectives.

Furthermore, the Strategy seeks to integrate more effectively ecological considerations into all relevant social and economic sectors, and will increase public participation in, and awareness and acceptance of, nature conservation interests. In this way, it intends to encourage a proactive approach by all the players involved. The Strategy encourages a more concerted and therefore more efficient use of existing policy, initiatives, mechanisms, funds, scientific research and information to maintain and enhance European biological and landscape diversity.

There are numerous players to be mobilised: national authorities, international organisations and financial institutions, organisations and associations active in the economic sectors, private enterprise, the research community, organisations responsible for disseminating information and educational establishments at every level.

Generally speaking, the Strategy sets out to achieve the following objectives:

- conservation, enhancement and restoration of key ecosystems, habitats, species and features of the landscape through the creation and effective management of the Pan-European Ecological Network;
- sustainable management of Europe's biological and landscape diversity through optimum use of the social and economic opportunities on a local, national and regional level;
- integration of biological and landscape diversity conservation and sustainable use objectives into all sectors managing or affecting such diversity;
- improved information on, and awareness of, biological and landscape diversity issues, and increased public participation in actions to conserve and enhance such diversity;
- improved understanding of the state of Europe's biological and landscape diversity and the processes that render them sustainable;
- assurance of adequate financial means to implement the Strategy.

The Strategy seeks results and aims to be more than just an expression of concepts but rather a dynamic process of preventative and reparative action. The Council and the Bureau for the Strategy are the organs responsible for its implementation. The Strategy's actions and achievements were presented to The Ministerial Conference on 'An Environment for Europe', held in Aarhus during 1998.

The Action Plan 1996-2000 identifies 11 eleven action themes, with Action Theme No. 0 being pan-European action to set up the Strategy process. Three of the themes concern issues requiring a co-ordinated pan-European approach (Action Themes 1 to 3), seven concern landscapes and ecosystems of significance (Action Themes 4 to 10) and one concerns endangered species (Action Theme 11):

- establishing a Pan-European Ecological Network;
- integration of biological and landscape diversity considerations into sectors;
- raising awareness and support with policy makers and the public;
- conservation of landscapes;
- coastal and marine ecosystems;
- river ecosystems and related wetlands;
- inland wetland ecosystems;
- grassland ecosystems;
- forest ecosystems;
- mountain ecosystems;
- action for threatened species.

The Council for the Strategy held its first meeting in Strasbourg from 15 to 17 May 1996. At that meeting it adopted a biennial work programme based on the Strategy's eleven action themes. The Bureau for the Strategy met from 27 to 29 November 1996 in Geneva and from 22 to 23 May 1997, in Strasbourg.

Action Theme 5 of the strategy, on coastal and marine ecosystems

The challenges to be addressed within Action Theme 5 are: direct loss through development and occupation of coastal areas for residential, touristic and industrial purposes, land reclamation, dams and dikes, coastal engineering, pollution, destruction and over-exploitation of benthic systems through industrial fishing practices, destruction of sedimentary systems through mining and drinking water production, and recreational disturbance. Certain pan-European objectives have been defined:

- develop and implement a European coastal and marine ecological network;
- develop an integrated coastal zone management approach to land and sea utilisation, management and planning as one system, based on conservation considerations;

- develop a special Coastal Code of Conduct, which provides clear recommendations and good practice rules to coastal authorities, project developers, coastal engineers and other user groups.

Other objectives have been defined at regional level:

- initiate priority conservation action for coastal and marine systems that play an important role in maintaining the biological and landscape diversity of the bio-geographical regions, focusing primarily on: northern Europe, the Atlantic, Boreal and Baltic regions and the Mediterranean;
- strengthen, establish and maintain priority conservation areas as key reproductive habitats for Monk seals and marine turtles in the eastern Mediterranean;
- establish an action plan for the conservation of Posidonia seagrass prairies in the Mediterranean;
- promote coastal tourism policies, that concentrate on quality improvement in existing resorts rather than on developing new ones in the Mediterranean and the Black Sea;
- assess methods to strengthen the position of bio-diversity and landscape assets in integrated models focusing on ensuring concentrated and compact rather than linear seafront development; focus on campaigns in the Mediterranean, Black Sea and Baltic Sea;
- establish action towards control and elimination of harmful exotic species in the Mediterranean and Black Sea;
- assist in joint conservation action for the Caspian Sea.

The Strategy's organs regarded this Action Theme to be a world priority and received special attention in the policies of the UNEP, the Council of Europe and the European Union. They appointed the UNEP responsible for its implementation. Also participating in its implementation are the United Nations Educational, Scientific and Cultural Organisation (UNESCO), the Council of Europe, the European Union, the Secretariats of the Barcelona Convention (for the protection of the Mediterranean Sea against pollution, 16 February 1976, as amended 15 June 1995), the Ramsar Convention (on wetlands of international importance especially as waterfowl habitat, 2 February 1971, as amended by the Protocol of 3 December 1982 and the amendments of 28 May 1987), the Bern Convention (on the conservation of European wildlife and natural habitats, 19 September 1979), the Bucharest Convention (on the protection of the Black Sea against pollution, 21 April 1992), the Helsinki Convention (for the protection of the marine environment of the Baltic Sea area, 9 April 1992), the European Union for Coastal Conservation (EUCC) and the Programme for the conservation of Arctic flora and fauna (AEPS).

At world level, it is worth remembering that Agenda 21 adopted in June 1992 in Rio de Janeiro to guide the action of the international community up to and during the next century, devotes its chapter 17 to the 'Protection of the oceans, all kinds of seas, including enclosed and semi-enclosed seas, and coastal areas and the protection, rational use and development of their living resources'. Since then, the UNEP has launched a programme for oceans and coastal areas, which has served to introduce action plans for many 'regional seas' throughout the world.

The Council of Europe, which has several times in the past addressed the issue of the protection of coastal areas - as we shall see below - has decided to help implement Action Theme 5. A Group of Specialists on coastal protection (PE-S-CO), set up in 1995 by a decision of the Committee of Ministers of the Council of Europe, met for the first time on 6 and 7 June 1996. The group noted that many technical and scientific activities on coastal protection had been carried out and numerous principles and legal texts had been drawn up. It found that all these activities highlighted the need for an integrated approach to coastal management and planning, but that despite all the efforts made so far, the state of coastal areas continued to deteriorate. The Group recognised that this resulted from the difficulties involved in implementing the concept of 'integrated management' and that it was now necessary to propose instruments facilitating the application of integrated management and planning, which were essential as operational instruments for the sustainable use of coastal areas.

The Group therefore proposed that the Council of Europe, in close co-operation with the EUCC and UNEP, launch action with two main objectives:

- to draft a code of conduct containing specific recommendations, practical realistic principles and good practice rules intended for local, regional and national authorities, project developers, coastal engineers and other user groups;
- to draft a model law on coastal protection defining the concept of integrated management and planning based on the principle of sustainable development, laying down the main principles to be

observed and making proposals regarding the appropriate institutions, procedures and instruments for implementing integrated planning and management. The model law could be used by the member States either to amend existing legislation or to adopt new legislation.

The Group proposed that the Council of Europe be instructed to draft the model law and that the EUCC draft the Code of Conduct in close co-operation with the Organisation.

This proposal was presented at a consultation meeting on Action Theme 5, organised by the UNEP on 26 and 27 September 1996, then approved by the Executive Bureau of the Strategy in November 1996. The first drafts for the model law and code of conduct will be examined by the Group of Specialists meeting on 26 and 27 June 1997. Work is still in the early stages but is being based on work already done by the Council of Europe on coastal areas in general, and on the Mediterranean and the Black Sea specifically.

Past activities and existing instruments

It was quickly observed that ill-considered development can be detrimental to the natural and cultural environment. Several studies published in the 'Nature and Environment' series, as well as meetings, warned of this danger and proposed solutions, which can be found in various resolutions and recommendations. Although these texts, adopted at international level, do not have binding legal force, we might legitimately ask whether, because of their quantity and similar content, from whatever angle the theme of 'environment-development' is approached (spatial planning, conservation, culture, major disasters, etc.), they do not already form a body of regional customary international law. Three international conventions on conservation of the natural and cultural heritage adopted by the Council of Europe - the Bern, Granada and Valletta Conventions - also have significant implications for coastal areas, even though they address broader issues.

Some of the texts deal specifically with coastal areas and maritime regions, while others, concerned with planning, spatial development and tourism, bio-diversity, natural habitats and protected areas, landscapes, cultural values and major disasters, are of great importance for coastal areas.

Texts on coastal areas and maritime regions

Resolution (73) 29 on the protection of coastal areas adopted by the Committee of Ministers on 26 October 1973, was already then aware that a considerable part of Europe's coasts was in a critical condition owing to the extremely serious biological degradation and aesthetic disfigurement caused by the indiscriminate siting of buildings, industry and tourist facilities in coastal areas. It noted that the situation was liable to deteriorate still further in the future, having regard on the one hand to the scarcity of coastal areas and the vulnerability of the seashore, and on the other to the increasing concentration of human activities in those areas. It therefore considered that protection of the coast could only be effective if multiple interests and problems were taken simultaneously into account, such as maintenance of the biological balance, preservation of the beauty of landscapes and conservation of natural resources, promotion of economic and tourist development and safeguarding of the hinterland.

Recommendation No. R (84) 2 of the Committee of Ministers to member States on the European Regional/Spatial Planning Charter, adopted on 25 January 1984 recommends that the governments of member States base their national policies on the principles and objectives set out in the European Regional/Spatial Planning Charter, which notes in its Appendix, on the subject of coastal areas and islands, that

> the development of mass tourism and transport in Europe and the industrialisation of coastal areas, islands and the sea, demand specific policies for these regions in order to ensure their balanced development and co-ordinated urbanisation, bearing in mind the requirements of environmental conservation and regional characteristics.

In addition, a number of discussions have been held as part of the activities of the Steering Committee for Regional Planning. From 7 to 9 May 1985, a European Seminar on the 'Development and Planning of Coastal Regions', held in Cuxhaven (FRG), produced some very useful proposals.

Recommendation No. R (85) 18 of the Committee of Ministers to member States concerning planning policies in maritime regions, adopted on 23 September 1985, emphasises that European maritime regions are at the same time a sensitive natural heritage and a particularly attractive area for mass tourism. It recommends that member State governments work towards the preparation of coastal planning strategies taking into account local, regional and national peculiarities of these areas as far as their economic, socio-cultural and environmental structures are concerned; encourage participation of regional and local authorities at the different stages of the preparation of such policies; promote at European level trans-frontier co-operation in the planning of coast and maritime zones and the hinterland; base their policies in this field on certain stated principles and objectives.

General texts referring to coastal areas

Planning, spatial development and tourism. Resolution 172 (1986) on tourism and environment, adopted by the Standing Conference of Local and Regional Authorities of Europe on 14 October 1986, aims among other things to limit and monitor all tourist activities which pollute the environment, to ensure strict use of land to prevent land speculation, and to assess the environmental impact of tourism projects. It also supports the promotion of all alternative forms of tourism which do not damage the environment and the safeguarding of natural areas of outstanding ecological interest. The Standing Conference of Local and Regional Authorities of Europe then held a Seminar on 1 and 2 September 1989 in Limassol (Cyprus) entitled 'Tourism and integrated planning policy', whose objective was to examine ways of encouraging environment-friendly tourism as part of integrated planning policies. The Final Declaration made several recommendations to local and regional authorities and was in favour of pursuing consideration of these issues.

Recommendation No. R ENV (90) 1 of the Committee of Ministers to member States on the European Conservation Strategy, adopted on 12 October 1990 at the 6th European Ministerial Conference on the Environment, recognises that environmental policies should be incorporated into all sectoral policies, including economic, social, cultural, educational, agricultural and forestry policies. The Strategy has a section devoted to 'seas', which states that it is necessary to

> avoid technical developments along the coastline, especially where it is still in a natural or semi-natural state, and protect coastal areas essential to marine fauna and flora.

The European Conference of Ministers responsible for Regional Planning (CEMAT) adopted a programme of activities focusing on forward-looking analysis of long-term, political, economic and ecological trends and developments. The programme is implemented through seminars that concentrate on finding a lasting, balanced solution to the various aspects of the many challenges for European society, facing all European states, as we approach the year 2000, and to which only an international consensus can produce a response.

The Colloquy entitled, 'The Challenges facing European society with the approach of the year 2000: strategies for sustainable quality tourism', held in Palermo on 4 and 5 June 1992, analysed the possible ways of striking a balance between the development of mass tourism and the need to protect the architectural and natural heritage. Conclusions were adopted for each of the three themes: balance between the development of mass tourism in vulnerable towns and natural areas and the requirement for the safeguarding of the architectural or natural heritage; the sustainable development of tourism in the Mediterranean: how to curb the unrestrained and sometimes irresponsible competition between tourist areas; and rural tourism as a socio-economic factor stabilising the rural population and an appropriate alternative to the pressure felt in overcrowded tourist areas. A further seminar organised as part of the CEMAT activities, was held in Athens from 25 to 27 April 1996 on 'The challenges facing European society with the approach of the year 2000: Strategies for the sustainable development of European states in the Mediterranean Basin'.

As far as tourism is concerned, it is also worth mentioning Recommendation No. R (94) 7 on a general policy for sustainable and environment-friendly tourism development, adopted by the Committee of Ministers on 5 September 1994. A draft recommendation on a policy for sustainable tourism in coastal areas is currently in preparation.

Biodiversity, natural habitats and protected areas. The Bern Convention on the conservation of European wildlife and natural habitats, adopted in Bern on 19 September 1979, aims to ensure that the conservation of wildlife and natural habitats is taken into consideration in general national policies on planning and development and in measures against pollution. In its preamble it recognises that wild fauna and flora constitute a natural heritage of aesthetic, scientific, cultural, recreational, economic and intrinsic value that needs to be preserved and handed on to future generations. The Contracting Parties stress the essential role played by wild fauna and flora in maintaining biological balances, and consider that their conservation should be taken into consideration by governments in their national goals and programmes and that international co-operation should be established to protect migratory species in particular.

The Convention currently has 36 Contracting parties, including 31

Council of Europe member States, 4 non-member States (Burkina Faso, Monaco, Senegal and Tunisia) and the European Community. It notes that numerous species of wild flora and fauna are being seriously depleted and that some of them are threatened with extinction and provides for a whole series of measures capable of rectifying this state of affairs if not reversing the trend that impoverishes bio-diversity. The legal obligations entered into by the Contracting Parties concern the protection of habitats and the conservation of species.

Several marine and coastal species enjoying strict protection are listed in the Appendices to the Convention. All the species of small cetaceans in the Mediterranean are therefore strictly protected, as are the four species of marine turtle *(Caretta caretta, Lepidochelys kern pu, Chelonia rnydas, Eretmochelys imbricata)* and the monk seal *(Monachus monachus)*. At its 15th and 16th meetings, held in January and December 1996, the Standing Committee for the Convention amended these Appendices to add a large number of Mediterranean marine and coastal species (see appendix below). The preparatory work was done jointly with the Regional Activity Centre for Specially Protected Areas, in Tunis (Mediterranean Action Plan), responsible for the implementation of the Geneva Protocol concerning Mediterranean specially protected areas of 3 April 1982 (henceforth the Protocol on specially protected areas and bio-diversity in the Mediterranean adopted in Barcelona on 10 June 1995). Since co-ordination, between the international conventions dealing with nature conservation is essential, the Standing Committee for the Bern Convention works with this in mind. In the 1980s, it began activities, which led to the adoption of the Monaco Agreement on the conservation of cetaceans of the Black Sea, the Mediterranean Sea and contiguous Atlantic areas.

The Standing Committee has also adopted several recommendations dealing with the protection of endangered species, coastal areas proper or the problem of the introduction of non- native marine species:

- Recommendation No. 6, adopted on 4 December 1986, on the protection of the Mediterranean monk seal (Monachus monachus);
- Recommendation No. 7, adopted on 11 December 1987, on the protection of marine turtles and their habitat;
- Recommendation No. 8, adopted on 11 December 1987, on the protection of marine turtles in Dalyan and other important areas in Turkey;

- Recommendation No. 9, adopted on 11 December 1987, on the protection of *Caretta caretta* in Laganas Bay, Zakynthos (Greece);
- Recommendation No. 12, adopted on 9 December 1988, concerning the protection of important turtle nesting beaches in Turkey;
- Recommendation No. 24, adopted on 11 January 1991, on the protection of some beaches in Turkey of particular importance to marine turtles;
- Recommendation No. 25, adopted on 6 December 1991, on the conservation of natural areas outside protected areas proper;
- Recommendation No. 45, adopted on 24 March 1995, on controlling the proliferation of *Caulerpa tax-folia* in the Mediterranean;
- Recommendation No. 54, adopted on 6 December 1996, on conservation of *Caretta caretta* at Patara (Turkey).

The Council of Europe has set up two networks of protected areas, including coastal sites. The authorities concerned are required to pursue their management in such a way that poses no risk to the environment.

The European network of bio-genetic reserves set up by Resolution (76) 17 of the Committee of Ministers of 15 March 1976, aims to maintain the biological balance and the effective conservation of the maximum number of representative examples of European flora, fauna and natural areas. There are currently 340 reserves in 23 States, several of which include a coastline. Resolution (92) 20 extended the network to European non-member States of the Council.

The European Diploma was set up in 1965 by Resolution (65) 6 of the Committee of Ministers for the purpose of protecting natural sites regarded as Europe's most prestigious on account of their international value and their scientific, cultural, aesthetic and/or recreational qualities. It is awarded for a renewable five-year period, and to date has been granted to 46 sites in 18 European States. The rules of the European Diploma are set out in Resolution (91) 16 adopted by the Committee of Ministers on 17 June 1991. Several of the prize-winning sites include coastal eco-systems: the Camargue national reserve (France), the Boschplaat nature reserve (Netherlands), the Minsmere nature reserve (United Kingdom), the Samaria national park (Greece), the Purbeck heritage coast (United Kingdom), the Doñana national park (Spain), the Scandola nature reserve (France), the Fair Isle scenic area (United Kingdom), the nature reserves of Bullerö and Langviksskar (Sweden), the Monte-Cristo

Island nature reserve (Italy), the Selvagen Islands nature reserve (Portugal), Maremma nature park (Italy) and the Tamniisaari Archipelago national park (Finland).

Landscapes. The Council of Europe has developed a field of activity on landscape conservation, which takes account of the impact that developments may have on the environment's aesthetic dimension. The Mediterranean Landscape Charter - the Charter of Seville, signed in Siena on 2 July 1993 by the regions of Andalusia, Languedoc-Roussillon and Tuscany, notes that the Mediterranean landscape is currently undergoing considerable changes as a result of the expansion of mass tourism and leisure activities which affect precisely those areas of the countryside which are of great social value. The Charter considers that a landscape conservation and management policy in the Mediterranean areas should target the following objectives:

- the conservation of areas of landscape with a particular historic or natural value representative of Mediterranean civilisations;
- ensuring that all human interventions are conducive to the creation of a landscape of the highest possible quality;
- ensuring that all development projects take account of the elements *in situ* which have natural, cultural or historical value;
- ensuring that schemes involving major transport, urban development, touristic or industrial infrastructures take due account of landscape preservation and, if necessary, make provision for rehabilitation;
- ensuring that all measures entailing the use or disposal of public property preserve the most important landscape areas having historical, cultural and natural value;
- ensuring that a fair balance is achieved between the landscape zones subject to restrictions and the neighbouring areas whose development benefits from their proximity to these zones.

Resolution 256 (1994) on the 3rd Conference of Mediterranean Regions (Taormina, Italy, 5-7April 1993) adopted by the CLRAE on 18 March 1994, subscribes to the aims of the Mediterranean Landscape Charter and invited the CLRAE to draw up, on the basis of that Charter, a framework Convention on the management and protection of the natural and cultural landscape of Europe as a whole. Work with a view to preparing a European landscape convention is currently in progress.

As far as cultural sites are concerned, Recommendation No. R (95) 9 of the Committee of Ministers to member States on the Integrated Conservation of Cultural Landscape Areas as part of Landscape Policies, adopted on 11 September 1995, considers the need to develop strategies for integrating landscape development and cultural site conservation into a global landscape policy, providing unified protection of the cultural, aesthetic, ecological, economic and social interests of the territory concerned.

Cultural assets. The Council of Europe has several activities in the field of cultural tourism, which often coincide with the concerns of environmentalists. It projects a modern image of the heritage, which is not limited to big historical monuments but covers all the components of the built environment and archaeology. Groups of specialists have put together a number of recommendations to the organisation's member States. In addition to the European Charter of the Architectural Heritage (1975) and Resolution (76) 28 concerning the adaptation of laws and regulations to the requirements of integrated conservation of the architectural heritage, there are also the Recommendations of the Committee of Ministers to member States on the protection of the twentieth century architectural heritage (R (91) 13), and on the protection of the architectural heritage against natural disasters (R (93) 9).

The European round tables which accompanied the launch of the European Plan for Archaeology, as part of the 'Heritages for Europe' from 19 to 21 September 1994 in Vienna, Bratislava and Budapest, noted that the concept of heritage ranged from historical monuments to cultural landscapes and the conceptual heritage. They noted that the transition from the architectural heritage to a cross-disciplinary cultural heritage policy involved the environment as well as activity sectors such as urban planning, regional planning, social policy and even politics in the broad sense, including the fight against intolerance and the protection of minorities.

Because they help to promote tourism, the European cultural routes contribute to the conservation of certain sites. In addition, a cultural tourism project was included in the Council for Cultural Co-operation's 1993 work programme. It aimed to help European citizens discover, through tourist activities, the treasures of Europe's cultural diversity and to bring such issues to the attention of both tourists and local, regional and national authorities.

The Granada Convention of 3 October 1985 for the Protection of the Architectural Heritage of Europe and the European Convention on the Protection of the Archaeological Heritage - the Valletta Convention - of 16

January 1992 both contain several provisions on 'integrated conservation' in which cultural property is considered in its environmental context. The Granada Convention provides that each Party shall undertake to adopt integrated conservation policies which:

- include the protection of the architectural heritage as an essential town and country planning objective and ensure that this requirement is taken into account at all stages both in the drawing up of development plans and in the procedures for authorising work;
- promote programmes for the restoration and maintenance of the architectural heritage;
- make the conservation, promotion and enhancement of the architectural heritage a major feature of cultural, environmental and planning policies;
- facilitate whenever possible in the town and country planning process the conservation and use of certain buildings whose intrinsic importance would not warrant protection but which are of interest from the point of view of their setting in the urban or rural environment and of the quality of life;
- foster, as being essential to the future of the architectural heritage, the application and development of traditional skills and materials.

Major risks. The Committee of Ministers of the Council of Europe concluded the EUR-OPA Major Hazards Agreement (Open Partial Agreement on the prevention of, protection against and organisation of relief in major natural and technological disasters) with the adoption in March 1987 of Resolution (87) 2. The aim is to stimulate co-operation between its member States by making use of current resources and knowledge to ensure more effective prevention of, protection against and organisation of relief in major natural and technological disasters (earthquakes, volcanic eruptions, industrial accidents, etc). The Agreement is open because any Council of Europe Member State can join, if accepted by the member States of the co-operation group set up.

A network of twelve specialist European centres conducts research and training in the fields of disaster-related medicine, protection of the cultural heritage, technological disasters, geodynamic and morphodynamic hazards, geodynamics in general, earthquakes, and accidental marine pollution. A European alert system in the event of a disaster is in place.

The European University Centre of the Cultural Heritage of Ravello, the

European Training Centre for Natural Disasters in Ankara, the European Centre on Prevention and Forecasting of Earthquakes in Athens, the Euro-Mediterranean Centre for Accidental Marine Pollution, Valletta, and the European Oceanology Observatory: forecasting major disasters and regeneration of the environment, in Monaco, are all important centres for the protection of coastal areas. The Mediterranean University, made up of a network of thirty or so universities in the Mediterranean region, has joined forces with a decentralised European training project on hazard sciences (Project FORM-OSE).

The Moscow Charter on the EUR-OPA Major Hazards Agreement, serving as a platform for co-operation between Eastern Europe, the southern Mediterranean and Western Europe in the field of major natural and technological disasters, adopted in Moscow on 2 October 1993, aims to organise rapid, effective relief operations.

The particular characteristics of the Mediterranean Sea and the Black Sea

The campaign for the protection of the Mediterranean coastlines. As a contribution to the conservation of what has been considered one of Europe's most magnificent regions, the Documentation and Information Centre for Environment and Nature (Centre NATUROPA), in 1988 launched a series of colloquies on the protection of the Mediterranean coasts, ending in 1991. Four meetings, in Messina (Italy) on 24 and 25 November 1988, Izmir (Turkey) on 19 and 20 October 1989, L'Escala (Spain) from 6 to 8 November 1990 and Bastia (France) from 30 May to 1 June 1991, looked at the main aspects of Mediterranean coastline protection through a study and a presentation of the various instruments capable of providing protection. Conclusions or recommendations were adopted at the close of each meeting.

The Colloquy in Messina looked at the issue of land purchase as a means of protecting coastal regions. The experiences of the *Conservatoire du littoral francais* and Britain's National Trust were analysed and presented as possible examples. The Izmir Colloquy highlighted the importance of spatial planning instruments as a means of protecting coasts. Development in depth was emphasised. The Colloquy in L'Escala looked at the instrument of 'protected areas' as a means of protecting the Mediterranean coasts, while the fourth Colloquy, in Bastia, dealt with marine parks as a means of achieving the same end.

More recently, the Centre NATUROPA organised four pan-European colloquies on tourism and the environment, one of which focused on the protection of deltas (Bucharest and the Danube delta, Romania, September 1992). The Final Declaration stressed the importance of an integrated spatial planning policy.

An international Colloquy on the protection of coastal areas of the Adriatic Sea, attended by many specialists on coastal problems, was held in September 1994 in Tirana, Albania, in advance of European Nature Conservation Year (ENCY) 1995. More than forty countries took part in this pan-European campaign which aimed to promote nature conservation not only within but also outside protected areas. The Conclusions adopted at the Tirana Colloquy note the need to draw up an international instrument containing specific undertakings and principles for coastal management and protection.

Conferences of Mediterranean regions and the Black Sea. Together with the Parliamentary Assembly, the Standing Conference of Local and Regional Authorities of Europe has been promoting interregional co-operation in the Mediterranean for 12 years. Four Conferences of Mediterranean Regions have been organised to date Marseilles (France) from 27 to 29 March 1985, Malaga (Spain) from 16 to 18 September 1987, Taormtina (Italy) from 5 to 7 April 1993, and Nicosia and Limassol (Cyprus) from 20 to 22 September 1995. The 5th Conference, to be held in the second half of 1998 in Montpellier (France), will look for the first time at the basins of both the Mediterranean and the Black Sea. These Conferences consider all aspects of environment issues.

The first Conferences of Mediterranean regions brought together participants from 16 Mediterranean countries to discuss problems of environment and development in the Mediterranean basin. The Marseilles Declaration adopted at the Conference - of which Resolution 162 (1985) on the Conference of the Regions of the Mediterranean Basin adopted on 16 October 1985 by the CLRAE took note - makes an evaluation of international co-operation, advocates better distribution of responsibilities between the State and local and regional authorities, facilitating regional consultation and makes a number of proposals (to disseminate international agreements on environmental protection among local and regional authorities, to involve those authorities in the measures outlined in the agreement, to set up permanent dialogue, to set up a department within each regional or local administration to be responsible for inter-regional or inter-municipal co-operation on environment protection, to develop regular contacts between leaders from the northern shores and southern shores of the Mediterranean, and so on).

Recommendation 1015 (1985) on protection of the Mediterranean environment, adopted on 25 September 1985 by the Parliamentary Assembly, asks the Committee of Ministers to define, within the general context of the North-South dialogue, a Mediterranean policy which should be concentrated initially on questions concerning inter-regional co-operation in matters of environmental protection, and include activities in the intergovernmental work programme designed to implement the Marseilles Declaration. Resolution 162 (1985) on the Conference of the regions of the Mediterranean Basin, adopted on 16 October 1985 by the CLRAE, makes proposals to promote interregional consultation.

The Malaga Conference had three main themes tourism in the Mediterranean basin, transport and communications in the Mediterranean basin, and ways of strengthening interregional co-operation. The Malaga Declaration recommends in particular that the Committee of Ministers examine the possibility of concluding a partial agreement between those Council of Europe member States concerned, open to accession by non-member Mediterranean states. The purpose of such an agreement would be to promote inter-regional co-operation in the Mediterranean Basin by co-ordinating the efforts of the various organisations, institutions or associations and taking appropriate initiatives in sectors not yet covered.

Referring to the Malaga Declaration, Resolution 200 (1989) on the 2nd Conference of Mediterranean Regions, adopted on 8 March 1989 by the CLRAE, encourages the local and regional authorities of Mediterranean countries to develop an area-based policy or sectoral policies (tourism, transport, etc.) aimed at reconciling development and environment conservation.

The Taormina Conference was a meeting of parliamentarians and representatives of regions and local authorities from all over Europe and from southern Mediterranean countries. The representatives of the Mediterranean countries adopted a Declaration stating that a policy for the Mediterranean regions must seek to achieve two objectives: firstly, to strengthen the position of the Mediterranean regions of the northern shore within the European Union and East-West co-operation, and secondly to place exchanges between the two shores of the Mediterranean in the context of North-South dialogue and co-operation. It also recommends strengthening inter-Mediterranean political and technical co-operation at all levels (central, regional and local) and defining a common strategy to achieve balanced economic growth throughout the region. The representatives of the Mediterranean regions also recommended that the Committee of Ministers of the Council of Europe adopt without delay the Convention on Inter-territorial Co-operation, whose draft was approved by the CLRAE in March 1993. This Convention, which is open to non-member

States, in particular those on the northern and southern shores of the Mediterranean, could serve as a legal basis for Mediterranean co-operation at local and regional level. The Declaration recommends that economic development be reconciled with the conservation and protection of natural resources based on a long-term approach. Special attention should be given to the following: joint management of water resources, protection of shores under threat from tourism, urban sprawl, industrialisation, transport and military activities, protection of the soil and action against pollution, and protection of the Mediterranean heritage (historic and cultural heritage, coastal landscapes, protected areas and nature parks). Resolution 256 (1994), mentioned above, on the 3rd Conference of Mediterranean Regions, adopted by the CLRAE on 18 March 1994, recommends that national and regional authorities and the Committee of Ministers take the various steps proposed in the Declaration.

The Conference held in Cyprus looked at the question of sustainable development as a fundamental strategy for dealing with demographic, migratory and environmental problems. The Final Declaration on 'sustainable development in the Mediterranean basin environment, demography and migrations' stresses the importance of co-operation among local and regional authorities for political stability and economic growth in the Mediterranean basin, and considers that the Council of Europe can help to promote and, where necessary, co-ordinate co-operation projects through, for example, the CLRAE. The Declaration dealt especially with demography, migration and action against intolerance, managing water resources and forest management.

Recommendation 21 (1996) on the 4th Conference of Mediterranean Regions (NicosiaLimassol, Cyprus, 20-22 September 1995) adopted on 2 July 1996 by the CLRAE recommends, among other things, that the Committee of Ministers adopt a Mediterranean policy consistent with the Council of Europe's traditions and functions, to be drawn up at governmental and parliamentary level and aimed at establishing specific programmes of co-operation complementary to those now being conducted for the central and east European countries; relating to respect for human rights, including the rights of minorities and equality between women and men, the rule of law, the establishment of democratic and pluralist institutions, demography and migration, the combating of racism and intolerance and the tackling of problems concerning the environment and the common cultural heritage); aimed at the territorial authorities on both sides of the Mediterranean Basin as the principal protagonists; co-ordinated the programme of action of the socio-cultural sector as adopted at the Euro-Mediterranean Conference held by the European Union in Barcelona on 27 and 28 November 1995.

37

At a hearing on co-operation in the Mediterranean Basin held in Thessalonika on 28 May 1996, the Parliamentary Assembly of the Council of Europe decided to concentrate its attention on interparliamentary and interregional co-operation.

At the first Inter-Parliamentary Conference on the Environmental Protection of the Black Sea was held from 10 to 12 July 1996 in Istanbul by the Parliamentary Assembly of the Council of Europe and the Parliamentary Assembly of the Black Sea Economic Co-operation (PABSEC). The participants asked that 1998 be proclaimed 'Year for the Conservation of the Mediterranean and Black Sea Systems' in order to create awareness at all levels of the problems of these two seas, to mobilise the ecological conscience of the peoples of European countries and of countries on the southern shore of the Mediterranean. The Parliamentary Assemblies of the Council of Europe and of PABSEC were invited to examine this proposal and adopt a position. These questions will be put before the Parliamentary Assembly of the Council of Europe in June 1997.

It has already been decided that the 5th Conference of Mediterranean Regions, to be held in Montpellier in the second half of 1998, will deal with both the Mediterranean Basin and the Black Sea. Interregional co-operation, peace, democratic security and sustainable development will be the focus of debates. The following have been invited to attend the preparatory meetings for the Conference as observers the PABSEC, the Inter-Parliamentary Union (IPU), the Arab Interparliamentary Union, the Assembly of European Regions (ARE), the Association of European Border Regions (AEBR), the Council of European Municipalities and Regions (CEMR), the Conference of Peripheral Maritime Regions (CPMR), the Organisation of Arab Towns and the Council of Europe's North-South Centre.

The participants in a seminar on incentive measures for the voluntary creation and management of protected areas organised as part of the programme of activities of the Bern Convention, in conjunction with the Ministry of Water, Forests and Environment of Romania, adopted in October 1996 the Constantza Declaration on the 'Year of the Conservation of the Mediterranean and Black Seas 1998'. Having taken note of the proposal to proclaim 1998 as the Year of the Conservation of the Mediterranean and Black Seas, the participants called for a common policy to improve the environmental situation of the integrated Black Sea-Mediterranean Sea system, including the Sea of Marmara, in the general interest of the Mediterranean and Black Sea populations, with a view to publicising at all levels the threats to the two seas, and encourage the active concern of the populations of all European countries and of the southern shore of the Mediterranean. They also wished that action be taken by the Bern Convention and Bucharest Convention in order to protect the coastal and marine bio-diversity of the Black Sea. In December 1996, the Standing Committee of the Bern Convention welcomed the Declaration with satisfaction and decided to take account of it in its Programme of Activities for 1998.

The trans-Mediterranean Programme (TransMed) was initiated by the European Centre for Global Interdependence and Solidarity - the North-South Centre of the Council of Europe, set up in Lisbon in 1990 - following a Symposium on Interdependence and Trans-Mediterranean Partnership, held in Rome in 1994, together with the Italian Government and the European Union. The purpose was to promote dialogue of Mediterranean partnership emphasising the interdependence of the political, economic and cultural dimensions. It contributes to enhancing awareness, communication and co-operation among non-governmental organisations, universities and research institutes, local and regional authorities, the media and youth organisations on the northern and southern shores of the Mediterranean.

Several activities were conducted to promote intercultural dialogue and deal with migration, youth and human rights.

A pilot project for an information service called Medgate, an information network on the Internet for organisations interested in Mediterranean issues, should serve to reinforce communication between institutions devising programmes concerned with the region, and develop complementary and synergistic approaches.

Conclusion

Knowledge of the ills afflicting coastal areas, just like knowledge of the remedies, if not complete is sufficiently advanced today. It is therefore the responsibility of governments and local leaders, as well as the 'consumers' of these areas, to ensure that their use does not destroy the very things that make them attractive. The coasts are remarkably rich in plant and animal species, particularly fascinating aesthetically but also especially fragile and under threat from all kinds of attack, and must therefore receive special attention.

The protection of the natural heritage belongs to a fast-growing awareness of the importance of environmental issues. As seas of navigation and trade, the Mediterranean and the Black Sea have for centuries played host to travellers interested in the trade, the religions, and also the beauty of the shores and the diversity of their resources. The use we make today of those natural riches must be sustainable. Economic growth, meaning human activities in general, must continue but the quality of the environment - which in the broad sense means the ecological processes and the various natural resources on which living organisms depend - must also be preserved. The survival of biological and landscape resources is essential for the needs and aspirations of future generations.

References

Augier, H., 1982. *Inventory and Classification of Marine Benthic Biocenoses of the Mediterranean,* Council of Europe Publications, Nature and Environment, 1982, No. 25, Strasbourg.

Augier, H., 1985. *Protected Marine Areas,* Council of Europe Publications, Nature and Environment, 1985, No. 31, Strasbourg.

Cognett, G., 1990. *Marine Reserves and Conservation of Mediterranean Coastal Habitats,* Council of Europe Publications, Nature and Environment, 1990, No. 50, Strasbourg.

Council of Europe, 1986. *Report of the European Seminar on the Development and Planning of Coastal Regions,* Cuxhaven, Study Series in European Regional Planning, No. 48.

Council of Europe, 1988. Proceedings of the 1st Colloquy *The Mediterranean Coasts and the Protection of the Environment,* Messina, Centre NATUROPA Publications.

Council of Europe, 1989. Proceedings of the 2nd Colloquy *The Mediterranean Coasts and the Protection of the Environment,* Izmir, Centre NATUROPA Publications.

Council of Europe, 1990. Proceedings of the 3rd Colloquy, *The Mediterranean Coasts and the Protection of the Environment,* L'Escala, Centre NATUROPA Publications.

Council of Europe, 1991(a). *Naturopa Special Issue on coastal zones of Europe,* No. 67.

Council of Europe, 1991(b). Proceedings of the 4th Colloquy, *The Mediterranean Coasts and the Protection of the Environment: Coasts and Marine Parks of the Mediterranean,* Bastia, Centre NATUROPA Publications.

Council of Europe, 1992(a). Proceedings of the 2nd Colloquy on *Tourist and Environment: Development and Safeguarding of Deltas,* Bucharest, Centre NATUROPA Publications.

Council of Europe, 1992(b). Proceedings of the Colloquy on *The challenges facing European Society with the Approach of the 21st Century: Strategies for Sustainable, Quality Tourism,* Council of Europe Publications, European Regional Planning, No. 53, Strasbourg.

Council of Europe, 1992(c). *Workshop on Management of Mediterranean Wetlands,*

Council of Europe Publications, Environmental Encounters, 1992, No.12. Strasbourg.

Council of Europe, 1993. *Second Workshop on Management of Mediterranean Wetlands,* Council of Europe, T-PVS (93) 4, Strasbourg.

Council of Europe, 1994(a). *Report of the International Symposium on Trans-Mediterranean Interdependence and Partnership,* Council of Europe Publications, North-South Centre.

Council of Europe, 1994(b). *Proceedings of the Workshop on Nature Conservation in Central and Eastern Europe: Present Situation, Needs and Role of the Bern Convention,* Council of Europe, Strasbourg, Environmental Encounters, 1994, No. 18.

Council of Europe, 1994(c). *European Heritage,* Cultural Heritage Division, No. 1.

Council of Europe, 1995(a). *Proceedings of the Intergovernmental Symposium on the United Nations Conference on Environment and Development (UNCED), the Convention on Biological Diversity and the Berne Convention: the Next Steps,* Council of Europe Publications, Strasbourg, Environmental Encounters, 1995, No. 22.

Council of Europe, 1995(b). *Proceedings of the Colloquy on the Protection of Coastal Areas of the Adriatic Sea,* Council of Europe Publications, Strasbourg, Environmental Encounters, 1995, No. 23.

Council of Europe, 1995(c). *Naturopa Special Issue on the Bern Convention,* No. 77.

Council of Europe, 1996. *Proceedings of the Seminar on the Challenges Facing European Society with the Approach of the Year 2000: Strategies for Sustainable Development of European States in the Mediterranean Basin,* Council of Europe Publications, European Regional Planning, No. 59.

Déjeant-Pons, M., 1987. *Protection et Développement du Bassin Méditerranéen - Textes et Document Internationaux,* Ed. Economica, Paris.

Déjeant-Pons, M. 1990. *La Méditerranée en Droit International de l'environnement,* Ed. Economica, Paris.

Stanners, D. & Bourdeau, P., 1995. *Europe's Environment,* Earthscan Publications, Luxembourg.

3 The limits of denationalisation and self-help in the areas of infrastructure and environmental protection

P. TRAPPE

Introduction

In this paper, I would like to discuss two current phenomena or two trends of our present historical moment that are in many ways complementary: denationalisation and self-help. The latter (i.e. self-help) is meant to awaken self-initiative in a socially and legally as yet unstructured, shapable sphere. We also speak of unlegislated areas, which are partly supported by those affected as they see fit, and partly by social groups taking the initiative with a competence based on experience (which is at once cost-effective and long-term). The relevant literature laments 'the missing institutional dimension' of environmental economics and practice.

In principle, the issue at stake is that the self-help sector should - or wants to - take charge of tasks that have so far been assumed by the state. Today, however, many no longer consider the state efficient enough. The present condition of state budgets makes a reduction of administrative expenses necessary; an expansion of the functions of public administration is in any case quite often beyond consideration.

In a historical perspective, the call for self-help has moved to the centre of socio-political interest once again. This is by no means impromptu, at least, in western European countries, and takes the shape of what is in principle a new appearance of social change. The real dimensions and the true potential of the capacity of self-help groups to date have perhaps not been recognised either by many citizens or by the specialised agencies of public administration.

Personally, I feel that we are at present experiencing an event of fundamental importance, namely the realignment of the areas of operation of the state and the people. The people are called upon increasingly to mobilise their own powers, the call being legitimated by what has been termed the 'privatisation of administrative functions'. What needs to be said right from the beginning is that both public administrations and independent social groups sponsoring self-help maintain that privatisation of this kind comes up against functional limits.

Self-help has a long and remarkable tradition in European social history, and its nineteenth-century conception continues to have an effect today. Developments that took off with the French Revolution and resulted in a strong increase of *voluntary associations* in England, France and Germany, in particular, serve to illustrate this tradition; it is illustrated, too, by the booming development of the co-operative sector in the nineteenth century, in particular of consumer co-operatives. Also, I should like to recall the economic and human plight of the increasing workforce under the conditions of early capitalist industrialisation. Chartism and the example of the *Rochdale Equitable Pioneers Society* come to mind as symptomatic cases in point. It is worth noting, furthermore, that as a rule real alternatives to private capitalism and free enterprise, such as production co-operatives, already faltered and got stuck in their trial period. Or take the case of 'New Harmony' in the United States, which Robert Owen conceived as a production co-operative and envisaged as a long-term relationship of its members (1825-1828). It is striking that the establishment of consumer co-operatives achieved remarkable widespread success and contributed significantly to public welfare despite rising real wages in late nineteenth-century western Europe.

Nineteenth-century consumer co-operatives gave rise to many contemporary and highly successful wholesale distributors, which continue to guarantee remarkably low prices for basic foodstuffs in particular. Here and there, these wholesalers have perhaps developed into joint-stock (i.e. public limited) companies out of the co-operative idea and co-operative structure. Yet it is noteworthy that their company name still bears the name of the co-operative, such as, for example, Coop Switzerland or the Migros Co-operative Association, whose memberships total around 1.2 million each - with a Swiss population of approximately 7 million (i.e. 1.2 million households, since membership is per household). Also, there are several networks of retailers that have joined together to form co-operative 'chains'. Bearing in mind that consumer co-operatives are not only open to members, but also encourage non-membership (the same is true of farmers' co-operatives, by the way), then roughly every Swiss household has a direct relationship with the consumer co-operative system.

The purpose of this reference to the consumer co-operative movement is to make clear that self-help organisations are by no means a novelty in European social history. In the following, I would like to establish the fact that a large number of self-help organisations are active in the areas of economic infrastructure and environmental protection, even though the strong trend of such organisations to expand their activities into the field of environmental issues is often overlooked.

Both Marx and some of his contemporaries, as well as a number of other thinkers, such as Karl August Wittfogel, Ferenc Tökey and Michael Atzler, who substantiated Marx's assessment in more depth later, have all drawn serious attention to the transition from Infrastructural self-help (i.e. the construction of roads, of canals, of coastal and inland water protection measures and, in particular, of drainage measures) to state control of such infrastructure measures and to the allegedly necessary passing on of self-help activities into state hands. Marx, for instance, remarked quite incisively that in 'ancient' and 'primitive' societies, the state apparatus took the 'public works' out of the hands of self-management. Infrastructural self-help, he observed, concerned the small social sphere, the subsistence economy of the clan or that of the political community. Marx might also be said to be suggesting that in view of increasing (social) differentiation, population growth and scarcity, only a central state agency could actually manage to administer or, rather, distribute public goods, such as water, in an extensive geographical area with adequate justice (i.e. doing justice to the traditional self-help units). In the social sphere, irrigation means that previously small autarchic units are joined together, which involves organisation and management at a higher, more complex level. This results in a new form of self-help activities and their organisation and/or the setting up of *specialised agencies* with state, or partly state-controlled, yet rarely private sponsorship.

When societies become more complex, it is necessary to establish a state organisation with a durable public administration, or what amounts to an apparatus that controls the beneficiaries of economic infrastructures (This involves obliging beneficiaries to contribute to maintaining the established infrastructure; also, it means committing them to paying charges for its use and benefit, that is to paying taxes). In the 1850s, Marx called this the system of the 'oriental means of production' on the basis of the British experience in India. He located the rise of centralised state rule, which involved a growing administrative apparatus, or what is known as bureaucratisation, in this inevitable change in the ways and means of organising infrastructure and production.

Other thinkers, such as Emile Durkheim, Ferdinand Tönnies, Georg Simmel, who have discussed the increase in social differentiation, have considered at most in passing the influences of state rule and public administration on the process of differentiation, respectively of what is involved when a social structure undergoes change under the circumstances of a developing large-scale society.

In a sociological view, the fundamental question about such tendencies towards change might be formulated as follows: how are we to explain the current thinking back to structures that have so far proved workable in small circles, in business and in self-management? How are we to explain this development, in particular, in the context of an unprecedented degree of social complexity and, what is more, under the joint influence of a reduction of national economic systems and increasing so-called globalisation? Such structures have called a number of competent social scientists into the arena of debate - mostly in view of the cuts in the social services in the late twentieth-century nation-state, a tendency that can no longer be passed over in silence. At this point, I should like to recall the efforts of communitarianism to establish new solidarities under the conditions of a complex, highly differentiated, 'economised' and highly mobile society - these conditions, it is worth adding, are quite unprecedented in the history of self-help movements.

These experiences, which empirical research clarified long ago, have meanwhile given rise to a particular field of inquiry for the social sciences, which is perhaps best captured by the notions of 'social capital' and/or 'social infrastructure'. The importance of a democratically institutionalised social infrastructure is clearly recognised in many areas of life. In this paper, I will focus expressly on the role of self-supporting processes and their assumption of possible functions in the economic infrastructure; in particular those geared towards measures in the environmental sector. Thus, what we are mainly concerned with here are the possibilities of self-help in the area of economic infrastructure. Such activities have manifold precursors, too, as well as already fully operational social infrastructures. Let me begin with a list of such activities in order to be in touch on at least areas in which self-help activities could be carried out without state support:

- co-operatives responsible for the upkeep of alpine paths;
- avalanche protection associations;
- associations responsible for the construction of barriers against landslides (by means of planting, and the construction of dams, hollows, and of other obstacles);

- independent irrigation and drainage associations, that join together from time to time in farmland co-operatives, particularly in the case of extensive developments or redevelopments of uncultivated land. I have come across interesting examples of water management under extreme conditions in Berber settlements in the Maghreb (particularly in southern Algeria) and in southern Arabia (Yemen, United Emirates, and Oman). In my view, however, the first (expensive) infrastructural measures (i.e. installation and maintenance) in these areas were not built through self-help;
- associations of owners of dyked land in coastal areas, which I have come across in northern Germany and Holland, in particular;
- forest protection associations that operate through voluntary fire brigade services (i.e. *de-facto* co-operatives, that are mostly organised on a municipal basis, and specialise in fighting forest fires, both in enforcing preventive measures and in taking action in emergencies);
- neighbourhood watch associations (endowed with police functions)
- 'surveillance associations in environmental law';
- engineering associations (i.e. associations of technicians and engineers), disaster control associations, technical relief services, emergency services (established on a permanent basis in Switzerland as part of the politics of development) etc.;
- last but not least, I would mention associations in the recycling sector that operate on the basis of waste disposal legislation, such as the increasing number of so-called waste disposal associations in Germany.

This list reminds us of the large number of self-help measures that can be taken. It is precisely in the present historical moment, in particular, that we are witnessing how such activities are spreading, with quite some ideological commitment and euphoria, to functions previously exercised exclusively by the state.

Quite plainly, functions and tasks of the state are passing into the hands of individuals and groups. This trend is supposed to grow even stronger, which explains why there is more and more talk of 'denationalisation'. In addition, there is a wave of privatisation that can be of advantage and disadvantage: Actual relief for the state is a definite advantage; the disadvantage, however, is a diminishing scope of action for those affected, together with a deteriorating material basis that would enable them to take action. Not only are denationalisation and the shift to self-help basically at issue here, but just as much the disincorporating (i.e. separating out) of essential resources from the traditional sphere of state action - of the nation-state's sphere of action - to processes that are trans and internationally interlinked, thus interdependent. Such processes are quite unknown in their present shape in European and world history, and they are able to elude the state's grip precisely because the ongoing process of privatisation stimulates them. It is thus neither very clever nor hardly surprising to state that the prospects for an increase in state revenue by way of taxation are hardly promising.

The processes I have described have limits, of course, in particular in those areas where the state is still expected to assume functions that could neither be taken over in a small circle by measures of self-help, nor by extensive integrated systems of self-help (such as global co-operative associations). As I am short of time, let me just mention some aspects of the process of denationalisation, respectively of privatisation, by raising a number of basic questions:

- who would assume the payment of state subsidies to regions lacking in infrastructure?
- who would assume the payment of state benefits to those on income support and to the growing number of those unable to take part in the arena of *global play?*
- who would conduct the supervision of important socio-political steering and of the bodies responsible for its operation?
- who would take charge of the guidelines for a presumably more complex and less clear domestic and foreign policy?
- and does the liability towards the environment not make a regularly constituted court of law necessary? Who can, or rather, who should enforce the principle of causation, i.e. the principle that the party responsible is liable for damages?
- what would happen to the classic responsibilities of the state, such as legislation, defence, tax policy and so forth? Surely, these areas could not be assumed by private or semi-official bodies, even if a well-meant amateurism considers this possible;
- who would keep in check the effects of private and self-help activities on third parties? We are familiar with such effects from the laws on industrial disputes in our labour legislations (I am

47

thinking of the effects of strikes on groups that are neither employers nor employees. In the social sciences, the case of specialised professional groups going on strike, such as pilots, air-traffic control staff, even cabin crew - such as at British Airways recently - is not considered a matter that must just be settled among those directly involved, but it is rather seen to involve considerably larger groups in industrial disputes at the very least as evidently injured parties, such as those who are cheated out of a service for which they have paid in advance).

We could, of course, start from the assumption that there are in fact areas of the law that open up enormous possibilities to self-help, such as the legislative framework for national (meanwhile also European) co-operative association laws. The question is rather to which extent citizens unattached to co-operatives (i.e. lacking co-operative ties), which probably includes the majority of humanity, want to come to terms with the economic and very much socially orientated decisions of those with co-operative ties. Basically, legislation would have to be expanded quite essentially (along lines similar to trade union laws), so that those unattached to co-operatives, but whose interests are nonetheless affected by the actions of those with co-operative ties, could already be taken into consideration by the law. Whether this is technically possible is beyond my judgement.

On the question of the wave of denationalisation, which can really and truly only amount to decentralisation, I should like to recall the famous model of workers' self-management in Yugoslavia, started already in 1948, as an example where a restructuring of the state, planned in line with such trends, was entirely unsuccessful. This was a case where there was extreme decentralisation and co-determination in all conceivable areas of the economy and the redimensioning of public administration that had been striven for at first. As a result, the self-administrative units set their own welfare first and foremost above general welfare; what is more, employees benefited in the first place from the profits of companies, with the issue put to the vote at the workforce's general meetings. State revenues declined sharply, so that responsibilities in the public sector could no longer be assumed sufficiently (such as the supply of necessary energy, with a functioning economic infrastructure, and the like); the state was hardly able to maintain its task of co-ordination. What proved disastrous for the former Yugoslavia, in particular for its economic development, was essentially that the reinvestment of profits on capital was neglected for the benefit of distributing operating earnings to the workforce, which after all believed it could decide the mode of distribution at its general meetings on the basis of self-management.

There was a general failure to recognise that such operations were connected to the public supply grid that virtually provided their economic-infrastructural basis. Instead of dealing with other examples of denationalisation processes that were conceived along the same lines in -but also outside - former socialist countries, I should like to consider a further structural problem, which might appear to be marginal in this context, that of decentralised judicial power.

In the business world, in particular, legal disputes are being settled increasingly by arbitration tribunals and their proposals of how judgement should be passed. In this way, certain disputes, mostly those that are very expensive, elude the course of justice of state jurisdiction. At first sight, this might seem to be a question of decentralisation, of an achievement of autonomy, indeed of self-help. In actual fact, however, it affects a principle that the western world has been striving for thousands of years, although it has never quite been achieved: the principle of the due course of law, which is affected inasmuch as the *rule of law* no longer prevails as the sole basis of the constitutional state, but is handed over to groups on the basis of declarations of legitimacy that have nothing at all to do with a democratically authorised social constitutional state. For their part, these groups practise legal interpretations they might even create law, but it is plain that this occurs without democratic control either being able to draw attention to abuse or being able to prevent it. Thus, we are talking about a side effect (and not merely as regards jurisdiction either) that has to do with social power, with group power, and with the power of group autonomies. Personally, I think it is safe to assume that every right and proper person would reject any suggestion that such a development is either desirable or should be carried to its limits regardless of the interest in denationalisation.

I should like to close by drawing at least some conclusions about the *limits* of self-help, which also concern the *limits* of the indispensable productive efficiency of external bodies (organisations, in particular semi-official or public institutions).

Let me begin with some hypotheses about self-help in the area of economic infrastructure (which includes the maintenance of an intact environment):

- if at all, self-help functions on a small scale that remains comprehensible for those it activates;

- self-help functions when objectives are clearly definable. Its ability to function decreases when objectives become increasingly complex;
- the ability of self-help to function decreases with the increasing complexity of the technical means required for self-help actions. I should add, in brackets, that integrating could increase the potential of self-help and co-ordinating various, even numerous self-help groups. This makes specialised functional agencies necessary, which are related to the actual process of self-help only indirectly;
- self-help is limited where external bearers of costs are required;
- in those cases where self-help comes up against the limiting factors of pursuing its objectives and has to hand over tasks to superior authorities (the case of bureaucratisation discussed by Marx and Max Weber), the self-help group can assume flanking, even perfectly essential, parts of the pursuance of objectives.

As for the social infrastructure, I would summarise the literature known to me and my own experiences as follows:

- Self-help functions in view of a clearly circumscribed objective - the self-help group forms by aligning itself with this objective.
- Accordingly, the degree of integration of a self-help group is low. High-level integration to the point of a long-term relationship or full-scale co-operative is atypical. Experience shows that it would have no reliable chance of survival: I refer to Owen's *New Harmony,* and to more contemporary experiences with collective farms (kolkhoz) and kibbutzim, as cases in point of fact.

Thus, it is not a question of personal or intimate, but rather of objectified or distanced social relationships. I am quite ready to acknowledge that the opinions of 'self-helpers' or their advocates are divided on this issue: it is not a matter of community-attachment, as the communitarians think. What is at stake is much rather the joint pursuit of aims, and thus the issue of solidarity, and not that of forced community building. Even though *social intimacy* is a viable proposition, it can seriously deflect the pursuit of aims.

Let me conclude with the thought that community for the sake of community exhausts itself and ends with the wish for social intimacy, which should not be confused with personal intimacy.

4 Regional and international conflicts within the coastal zone. A case for partnerships and European-wide cooperation

R.W. DIXON-GOUGH

Definition of the coastal zone and coastal zone management

Throughout this paper, the terms *coastal zone* and *Coastal Zone Management* are used extensively. In order to avoid confusion, these terms are defined below.

The coastal zone

The coastal zone is taken to include the rivers and tidal inlets, the terrestrial coast (sea cliffs, sand dunes, shingle structures and coastal grazing marshes), inter-tidal shores (including sandy beaches, mudflats and rocky shores) and the marine zone to the offshore limit of coastal processes, together with the plants and animals living within the zone. Geomorphological processes form the basis for the development of coastal habitats and the wildlife, which they support. These may operate over a wide geographical area and essentially provides the link between the land and the body of water.

Superimposed in this is the factor of mobility, not only because the zone is dynamic but also of the many species of animals, notably migrating birds, sea mammals and fish, which are dependent upon this zone. Human impact in this zone can be equally wide-ranging and the potential scale and complexity of the interactions can be immense. For planning purposes, the zone can be defined according to the human uses and activities, which may take place there. Taken from both perspectives, the zone is defined as a combination of natural features and human activities, which may interact across the whole zone or within individual components of the zone. Recognition of this interaction is the first stage in understanding the need for an integrated approach to management of the zone.

Coastal Zone Management (CZM)

From the point of view of nature conservation, CZM could be defined as a means of providing a framework for the development of integrated strategies for the protection of the natural coastline and marine areas, including the dynamic operation of coastal processes. However, by this definition, the importance of human use and exploitation of the zone and the impact of its ability to accommodate change without any loss of interest, is ignored. Thus a more pragmatic definition might be:

> Coastal Zone Management provides a mechanism for the integration of human activities within a zone defined by natural processes which facilitates the sustained use and exploitation without degrading the environment (Doody, 1995).

Introduction

Many factors have an important influence upon the coastal zone and these effects are truly transnational. It is primarily for this reason, that coastal zone management is assuming such importance throughout the world. Dixon-Gough *et al.*(1998) discusses the global environmental problems, one of which was deforestation. Deforestation, however, not only affects the land but also has potentially serious consequences for adjacent regions. For example, deforestation in Jamaica has lead to severe soil erosion and an increase in suspended sediment loads in many of the rivers. These, in turn, drain into the surrounding coastal regions with fringing coral reefs (Hughes, 1994). The corals are very sensitive to concentrations of suspended sediment and, as a result of reduced light penetration, their rate of growth is slowed. Additionally, the concentrations of sediment poison the tissues of the coral, resulting in a degradation of Jamaica's coral reefs. As these begin to die, this in turn affects other life forms, which depend upon the reef for food and which are part of the coral reef ecosystem.

In many respects, and particularly when taken in its traditional context, coastal zone management has been largely a story of coastal mismanagement, principally because of localised influences. During the last two centuries or more, the European coastline has been under the twin threats of urban development and land reclamation. The largely uncontrolled urban development stems initially from the need to build and develop industries close to coastal and international trading routes and the infrastructure needed to support those industries. Land reclaimed from the sea is also an extensive feature in most European countries and that land may be used for a variety of purposes, ranging from the provision of land for industrial development to making more land available for agriculture. As a consequence, the land must be protected from the natural action of the sea. Many cities, both in Europe and throughout the World, are located directly on the coast, while others are within easy reach of it. Additionally, the growing popularity of coastal resort holidays brings further population pressure to bear on many coastal areas (Healy, 1995).

In Europe, engineers developed the skills to provide appropriate sea defences, both to prevent flooding caused by high tides and surges (but also to reduce coastal erosion) whilst simultaneously developing the infrastructure of the land behind the sea defences. Along the coastline of Britain, the sea defences have been constructed with the 'fifty-year' storm in mind. This has led to the building of sea defences that have often obscured the views from the properties that they have been designed to protect. This negated the very reason for the existence of the properties, which are frequently named because of their scenic positions, two common names being 'Belle Vista' and 'Sea View', but regrettably - no more! In Britain alone, approximately 50 per cent of the coastline are now dominated by man-made structures. Natural features, such as salt marshes and dunes are virtually limited to coastal natural reserves (Clayton, 1993).

Unfortunately, many of the schemes for coastal protection were considered by civil engineers to be appropriate treatment for perceived littoral problems, particularly those of flood prevention and the stabilisation of existing coasts. Furthermore, those schemes have largely been undertaken as part of a practical process following several natural problems. Through studies of old maps (some as recent as 1950) it is possible to identify how the coastlines of Europe have changed.

When considering the process of CZM, it is essentials to recognise that many features exist that have (often in conflict with one another) factors, which are worthy of attention. The most important of these is the realisation and understanding that the coast - whether it is a hard or soft coastline - is a dynamic region, which must be taken into consideration when attempting to preserve, reinstate or reinforce portions of the coast. Secondly, the very factors which shape the coastline, changing patterns of wave energy, changes in sea level, varying supplies of sediment, together with coastal geology and geomorphology constitute natural hazards against which it might be deemed necessary to protect the coastline and those living within the coastal zone.

Finally, the coastline forms a recreational region far more popular than the majority of inland regions with the possible exception of National Parks (Dixon-Gough *et al.,* 1998). This factor has led to a number of measures being undertaken to cater for tourism, such as the creation of artificial beaches, beach retention schemes, as well as the associated patterns of pollution.

The large, settled and transient coastal population requires the support of an industrial, commercial, agricultural and service infrastructure. As a result, coastal areas - particularly coastal lowlands - are often the hubs of intense human activity. Coastal lowlands are usually the most fragile coastal environments, with complex feedback exchange interactions with the sea (Tooley, 1987). The popularity of these areas with human beings is mirrored by the great diversity of flora and fauna in coastal regions. A wide range of habitats make up several coastal bio-geographical regions in Europe (Huggett, 1995).

A framework for coastal zone management

It is essential that the range of actual and potential conflict between nature conservation, industry, tourism and recreation is explored and evaluated within a CZM framework. Although these factors are primarily concerned with the coastline, other considerations also relate to the coastal zone. For example, Europe has a high percentage of nations, which have traditionally been involved in sea trading. Those nations require ports and port facilities. Normally associated with the ports are industries dependant upon imported raw materials, typified by the steel industry in South Wales and the oil refineries of Rotterdam. Other industries, such as the generation of electricity require access to large quantities of cooling water and are consequently often located within the coastal zone. Thus, conflicts are created between the 'natural' environment, the requirements and effects of industry, and the special requirements of recreation and tourism. Finally, there are those coastal regions still dedicated to agriculture, much of which is protected as sites of natural beauty or designated as various types of nature reserves.

The concept of coastal zone management has arisen from recognition of the need for coastal management to do better than merely defending the shoreline. Selective strategic planning, involving a wide range of coastal activities and interests, is generally accepted as an approach which is preferable to 'crisis response'.

Far too often, CZM is reactionary rather than planned. Perhaps this is being unfair since it is rarely possible to plan for the future without the knowledge of past and existing experiences, thereby allowing the management process to be based upon a firm foundation of scientific understanding and previous experience. However, due to the fragmentation of responsibility for CZM, particularly within the UK, the planning and management of the coastal zone has traditionally been based upon localised plans, often developed rapidly as a reaction to a perceived or existing problem. In these cases, very little consideration can be given towards its effect upon adjacent areas.

Ketchum (1972) lists six major spheres of anthropogenic activities that are of particular relevance to the coast and that are likely to influence the development of the framework for a coastal zone management structure. These are:

- residency and recreation;
- industry and commerce;
- waste disposal;
- agriculture, aquaculture and fishing;
- conservation and the presence of nature reserves;
- military and strategic functions.

In an ideal situation, when there is sufficient space for all the above activities to function within defined spheres or sectors, there would be little competition for space. The pressure that is generated within the coastal zone is normally the result of overlapping interests and sectors, which combine to compete with the 'natural environment' of the coastal region. Whilst careful planning and the designation of conservation and environmentally sensitive areas might reduce conflict within certain defined sectors, some activity matrices in some areas are becoming extremely complex (Ruddle *et al.*, 1988).

Carter (1988) identifies three clear approaches to CZM:

1. an undemanding 'Friday afternoon' task whereby 'off-the-shelf' solutions are applied to problems ranging from coastal erosion through to the conservation of habitats;
2. an increased appreciation of coastal issues, although the sheer magnitude of many of the tasks, embracing a wide range of impacts often operating across different time-scales, may be off-putting;
3. a deeper understanding, in which coasts are viewed as dynamic systems, which link together physical, chemical, biological and

socio-economic processes.

The latter approach can be further refined for management purposes into a series of sub-systems which each cover a range of time-scales and areas. For example, a coast would incorporate basic concepts such as energy, the mobility of sediments, nutrient exchange and resources. Thus, a defined area could be assessed dynamically within a sub-system yet also against adjoining systems relating to the shore (inshore waters, dunes, marshes, cliffs and other geomorphic features), habitats and to its associated socio-economic values. The latter would include factors such as coastal protection, fisheries, tourism and recreation, urban and industrial infrastructure, and development potential. This type of approach allows for the generation of models to assess the effects and the capability of the system to absorb pressure without irreversible changes. Clearly, coastal zone management systems must integrate data from a variety of sources.

Once the modelling has been successfully achieved for individual sub-systems, they may be combined and integrated at a regional scale to a consistent approach. The main problem concerning the integrated approach lies in the existing administrative divisions, which inevitably lead to the artificial segmentation of the shoreline, which has little resemblance to natural boundaries. A further problem relates to the management systems for which there is rarely a lead organisation. This can lead either to conflict, or simply to a lack of action. Private ownership can create other problems, particularly when the owners cannot afford a reasonable and consistent standard of management. These factors point towards the development of a coherent and integrative framework of coastal zone management, which will certainly extend across regional and international boundaries throughout Europe. Although no such CZM framework exists across international boundaries, it is worth considering the requirements of such a scheme and examples of the extensive CZM programmes elsewhere in the world. The process of coastal zone management thus develops into a strategy of balancing the spheres of human activity.

The many activities of the coast imperil, or are imperilled by, the growing demands placed upon the coast (Healy, 1995). The increases in population along the coast gives rise to concerns about the quality of the coastal resource, including shoreline erosion and sea defence, habitat degradation, pollution and the reduction of coastal diversity as a result of ever-increasing pressure on a dynamic and often fragile environment.

Demands within the coastal zone are increasing at a time when their resource base is shrinking. Many of the demands are derived from an identification of previously unrecognised issues, such as the need to protect natural habitats and other environmental considerations. This growing web of environmental linkages is proving to be difficult to resolve for most planners and managers associated with the coast. To assist in this process, it is essential that a framework be developed - both to provide consistency and to ease the problems of communication. Carter (*op cit*) identifies two ways in which this might be achieved:

1. by imposing a new structure across existing areas of responsibility, such as the Coastal Zone Management Act in the USA, and
2. by developing a framework based upon existing groups. In the best examples of this category, a lead agency is designated through which all other groups can co-ordinate their activities; whilst in the worst examples, responsibility is apportioned according to existing criteria with little opportunity provided for overview (Carter, 1988).

In both types of framework, there should be a defined association between a coastal policy and a coastal plan in which the economic and socio-economic issues are considered, whilst trends and indicators of change might be identified. It is only possible to achieve international, national and regional consistency if policies and plans are clearly defined and adhered to whilst allowing local implementation to take into account site factors and the distinctive characteristics of individual regions.

The political dimension

Because of its transnational nature, it is important to seek regional, international and even global recognition of the problem. This approach, however, is fraught with difficulties since most territorial waters are deemed to have an economic benefit to countries involved, many of which jealously guard their territory. There have, however, been two note-worthy events, which indicate that a spirit of co-operation could lead us into the twenty-first century with some degree of confidence for global and regional coastal zone management. These are, the Earth Summit, held at Rio de Janeiro in June 1992 and the United Nations Regional Seas Programme.

The Earth Summit

One of the most influential events in recent years, which has helped place coastal management high on the political agenda, was the UN Conference on the Environment and Development (the Earth Summit) held in Rio de Janeiro in June 1992. One of the most important aspects of the conference was the reconciliation of the twin goals of economic growth and environmental protection under the general political focus of "Sustainable Use". In this context, Agenda 21 identified the need for a comprehensive programme of action to achieve a sustainable pattern of development over the next 100 years. Chapter 17 of Agenda 21 specifically deals with the

> protection of the oceans, all kinds of seas, including enclosed and semi-enclosed seas, and coastal areas and the protection, rational use and development of living resources.

Programme Area A, which is specifically concerned with the integrated management and sustainable development of coastal and marine areas, includes the provision of data and information as one of its three activities. In particular, it states that

> information for management purposes should receive priority, supporting the view of the intensity and magnitude of the changes occurring in the coastal and marine areas.

Each of the other programmes also identifies the provision of information as being of key importance. Programme F, which deals with international and regional co-operation, specifically identifies as a key issue

> strengthening the capacity of international organisations to handle information and support the development of national, sub-regional and regional data and information systems

Chapter 15, on the conservation of biological diversity, also identifies the provision of data and information as a primary activity. It specifically suggests the establishment of

> baseline information on biological resources, including coastal and marine ecosystems, as well as inventories undertaken with the participation of local communities.

This latter point is of great significance, since in many areas, the implementation of a more enlightened exploitation of the coastal zone can only be achieved with the active participation and understanding of those who live and work there.

The Regional Seas Programme

The United Nations Environmental Programme (UNEP) has become a leading force in the development of international environmental programmes. One such programme addresses the management of the marine environment through its Regional Seas Programme. The role of UNEP in ocean management was initiated in 1974 - two years after the establishment of UNEP - at the Stockholm Conference of the Human Environment, when oceans were designated for priority management efforts. A regional approach to the control of regional pollution and the management of marine resources was endorsed and this was initiated as the Regional Seas Programme (UNEP, 1991).

This programme was given impetus at the 1976 Barcelona Conference for the Protection of the Mediterranean Sea Against Pollution (UNEP, 1982) which placed the emphasis upon regional water quality planning and marine resource management. In 1982, a further report was commissioned by UNEP which focused the programme on coastal rather than open ocean pollution programmes (UNEP, 1982a).

The basic model for UNEP's international marine environment planning is the development of comprehensive regional programmes for marine environment conservation together with a parallel action plan. Although it often been assumed that the first convention and action plan was for the Mediterranean, others contended that the 1974 Helsinki Convention on the Baltic Sea. This addressed the numerous pollution sources and emphasised the need for scientific understanding and joint decision making and was pivotal in the development of the Regional Seas programme (Hulm, 1983). The historical influence of such earlier 'Northern' agreements by industrialised countries was acknowledged by Sand (1988) but he also noted the critical elements that make the Barcelona Convention the first of its kind incorporating developing countries into an action plan for regional environmental planning and development. The significant role played by the Food and Agriculture Organisation of the UN (FAO), which focused the initial concerns about the pollution of the Mediterranean is noted by Haas (1990). The control of the pollution issue was subsequently transferred from FAO to UNEP (Miles, 1983). The Mediterranean Action Plan (MedPlan) has been considered as the template from which all other action plans have been developed (Alexander, 1990; Vallega, 1992).

The Regional Seas Programme has now been used to provide a relatively consistent format for 12 regions, each action plan containing five basic elements; environmental assessment, environmental management, legal requirements, institutional arrangements and financial arrangements. Optional or sub-regional measures may also be added to the action plans and these have included co-ordination, support measures (such as public awareness), and the development of human resources. Whilst pollution issues are the primary focus of the regional conventions, a broad array of other issues include coastal habitats and resource management.

In keeping with the importance of pollution issues, environmental assessment is considered to be the priority activity of most of the action plans. The main goal of the environmental assessment is to establish links between assessment activities and environmental management decisions by allowing

> national policy makers to manage their natural resources in a more effective and sustainable manner and to provide information on the effectiveness of legal/administrative measures taken to improve the quality of the environment. UNEP, 1991).

The emphasis is on assessing and evaluating the causes, magnitude, and consequences of environmental problems (UNEP, 1982a). Specific elements included in the environmental assessment might be studies in environmental baseline conditions, contribution of pollution sources, and socio-economic conditions affecting the coastal or ocean environment. The environmental management section of the plans focuses on the management programmes and structures that exist, or need to be developed, to use the environmental assessment data in managing environmental resources. Typical programmes include Environmental Impact Assessment (EIA), CZM and emergency contingency planning.

The legal section of the action plan sets out the elements of the regional framework convention and supporting technical protocols. The actual conventions and protocols are adopted separately and must be ratified by the individual countries. The section on institutional arrangements relates to the development of institutional mechanisms to ensure that the action plan is implemented. This is explicitly recognised to ensure that progress of the action plan is maintained, whilst enabling a mechanism to allow the approval of new activities and financial support (UNEP, 1991). This includes the involvement of a range of organisations; regional bodies, national focal points, sub-regional and international institutions.

The RSP currently includes 12 existing action plans involving 130 states, 15 bodies of the United Nations, and 13 other international organisations (Sand, 1988; UNEP, 1991; Siren, 1992c). The programme has resulted in the development and support of the following conventions and protocols:

- 9 marine pollution conventions;
- 9 protocols for co-operation in combating pollution from oil and harmful substances;
- 4 protocols for controlling land-based pollution sources;
- 3 protocols for dumping of hazardous materials;
- 4 protocols relating to the conservation of protected marine and coastal areas, and wildlife;
- 1 protocol each on pollution from continental shelf exploration and radioactive waste disposal.

The best measure of the success of any environmental programme should be based upon measurements made in situ, i.e., outcome evaluation (Jacobson, 1995). Both technical and non-technical criteria could be used to determine environmental improvements. Examples might include; improved beneficial uses; decreased contaminant concentrations and flows from industrial and municipal facilities, higher standing stocks of marine organisms; lower contaminant levels in water, sediments and fish, together with habitat improvements or protection (e.g., the creation of reserves and wetland restoration). Providing that sufficient and adequate information could be compiled from each of the RSPs, a programme-wide outcome analysis might be accomplished. It is far more likely, however, that one specific RSP that has developed a regional monitoring programme would be the best choice for such a detailed evaluation. Even for such a well-researched and documented area as the Mediterranean, no such compilation exists (Dobbin, 1993). Although some work has commenced on compiling information on several databases for the Mediterranean, there is no single, comprehensive database containing appropriate data that could be used to examine the long-term quality trends (Haas, 1990). Problems regarding the appropriateness of the data are largely the result of inadequate spatial and temporal coverage.

Regional assessment efforts, however, have a good reputation for local support and have resulted in numerous technical and country-specific reports;

75 for the Mediterranean, 54 for the South Pacific region, and 18 for the Caribbean (UNEP, 1993). One such assessment, conducted in the Caribbean, documented the major land-based sources of pollution for the island states (Hinrichsen, 1990). However, many individual states within regions lack the expertise, financial and technical resources to conduct sophisticated monitoring exercises. Even though the countries in the Mediterranean region have established an extensive monitoring programme, (MEDPOL) that has succeeded in focusing the environmental priorities for the region, continuing problems such as the lack of representative data, inadequate quality control for the data and the completion of only as few co-operative studies. Despite having the most extensive monitoring programme of any RSP, a recent evaluation MEDPOL found that there were serious problems with the monitoring programme. The evaluation concluded that

> the results....still cannot provide the contracting parties with a description of the state of the marine environment in the Mediterranean, or a provisional estimate of the contribution of inputs to the Mediterranean or an indication of temporal contaminants in marine organisms. (Siren, 1993a).

For states within regions that lack the sophisticated scientific infrastructure and capabilities of the Mediterranean region, concern with more basic issues takes precedence over making environmental assessment the main thrust of management structures and activities. For most regions within the RSP, information on pollution sources other than dumping and accidental spills (from land based sources) and comprehensive delineation and analyses of the major environmental problems are still lacking (UNEP, 1982a; Hulm, 1983). The development of adequate monitoring programmes to assess environmental problems and to provide input to environmental management decisions has yet to be realised for most regions. Even when monitoring programmes are developed, regional co-ordination is sometimes lacking.

Every action plan lists some region-specific environmental management goals. Examples include pollution control, CZM, and fisheries protection. The RSP has served to focus and strengthen marine pollution control efforts in all areas. Inter-regional training and adoption of environmental management programmes has improved existing marine pollution control programmes. Additional sewage treatment plants and oil reception facilities have been built in the Mediterranean (Haas, 1991; Hinrichsen, 1990).

International legal instruments for coastal protection

There is little chance of harmonising the legislation for coastal management on an international, let alone a European level. Natural and political circumstances have, so far, prevented this. However, a minimum standard for coastal conservation is needed. Since the beginning of 1992 the European Union has been working on this but no European legislation seems to be in sight in the near future.

As coastal conservation is a pan European issue, other European structures may be investigated as offering a potential way forward. As a result of the Helsinki Convention (Convention for the Protection of the Marine Environment of the Baltic Sea Area), there are now provisions concerning nature conservation and biodiversity (Article 15). More precise rules than those in the Helsinki Convention would be needed. In principle, the Bern Convention is suitable for this, although until now it has been concerned mainly with the protection of individual species; coastal conservation and landscape protection are not considered. Another alternative is to initiate an entirely new convention on European coasts. The main questions that such a convention should deal with includes:

- a definition of general principles for the use of the coast;
- minimum standards for a protection zone along the coastline outside already urbanised areas and planning zonation regulations.

It must be remembered that the European coast is the common heritage of a wider Europe. It contains some of the wildest and least spoilt areas, together with some of the most commercialised and polluted areas of Europe. In many regions, little can be done to remedy the mistakes of the past. It is, however, very important that what exists that is good is conserved and that that is spoilt is improved upon. This generation is responsible for acting as custodians for subsequent generations so that the extraordinary beauty of the coastal landscape is preserved and the biological importance of coastal ecosystems for the wild fauna and flora on our continent can be maintained

European coastal zone management

The European Environment Commissioner, Ritt Bjerregaard, warned in 1996 that the EU must take strong action if it wishes to safeguard its wetlands. In a

Commission Communication ('Wise use and conservation of wetlands' - *Com doc(95) 189 final*), she pointed out that more than half of Europe's wetlands have disappeared in recent years as a result of excessive urbanisation and unsound agricultural practices (Anon, 1996a).

CZM has been on both national and EU political agendas now for some 25 years (Huggett, 1996). As early as 1970, changes to the way in which the coastline of England and Wales was managed were recommended (Countryside Commission, 1970). During this period, the Council of Europe began promoting integrated coastal planning to help ensure wildlife conservation in coastal areas. Despite this awareness, coastal habitats continued to be degraded, largely because action was being focused at a local level whilst many of the issues needing to be addressed were trans-boundary. These include navigation and communication, waste disposal and pollution control, nature and landscape conservation, recreation and tourism, fisheries, minerals and energy exploitation, and coastal engineering (Gubbay, 1994). In response to these concerns, the Council of Environment Ministers adopted a resolution calling for the production of an EU strategy for integrated CZM and its incorporation into the 5th Environmental Action Programme (CEC, 1992).

Progress has been very slow and in 1994, the Council of the European Union renewed its invitation for the Commission to develop a strategy for ICZM for the entire coastline of Member States. In 1995, the Commission published a Communication on CZM, which proposed the establishment of a demonstration programme (CEC, 1995; 1996). The aims were to:

- show how to increase the effectiveness of existing instruments through the principles of integration and subsidiarity;
- show how to achieve better co-ordination between sectors across all levels (Europe through to local levels);
- test mechanisms of co-operation.

In the report, the Commission concluded that there is a justification for Community action with regard to the development of a coastal zone programme. The three reasons given for this interest are:

1. the existence of problems of a European dimension which cannot be solved by the separate countries;
2. the influence of EU policies and action on the development of a coastal zone;

3. the need for an exchange of experience and know-how in a field where successes are rare and where there is substantial public and political demand for the conservation of the coastal zones and their sustainability.

One approach, identified by the European Environment Agency, contains the following recommendations (da Silva, 1996):

- identify the main aims and objectives for CZM throughout the EU;
- review EU Directives and policies with a potential impact on activities in the coastal zone;
- identify the relative importance of these instruments for the state of the environment of the coastal zone in the long term;
- assess those instruments through which the CZM strategy could be structured
- identify the mechanisms needed to co-ordinate the development and implementation of coastal zone policies, both within the Commission and the Member States;
- identify means to assess progress in reaching the objectives of CZM;
- identify actions that would be required by the Commission and by Member States to implement the strategy, including the financial implications.

These recommendations were of particular significance since they appear to reiterate the desire of Chapter 15, of Agenda 21 of the Earth Summit in which the responsibility for environmental care should incorporate local responsibility.

European integrated management programme

Initially, one must pose the question, 'What is Europe?' In a variety of stated measurements, Europe's coastline variously extends from approximately 89,000 km, to over 150,999 km. Irrespective of the absolute length of the coastline, it is important both in socio-economic terms (47% of the EU's population live within 50 km of it) and in nature conservation terms. Much of the nature is, however, under serious threat. In recognition of the European dimension of this problem, the Commission proposed in 1995 to launch a three-year demonstration programme on the integrated management of its coastal zones. This aims specifically at encouraging cross border and cross sectorial dialogue and concerted action, so as to facilitate the implementation of existing legislation in this area (Anon, 1996b). Th general principles and policy options and lessons learnt from the demonstrations programme, were eventually published in two volumes (EC, 1999a; 1999b).

The European Union (EU) adopted the Habitats Directive (HD) 92/43/EEC on the conservation of natural habitats and of wild fauna and flora in May 1992. The main aim of the HD is to promote the maintenance of biodiversity, but taking into account the economic, social and cultural requirements of the regions. The HD provides for the establishment of a network of protected sites across Europe which will be known as 'Natura 2000'. This network will include sites designated as Special Areas of Conservation (SACs) under the HD and Special Protected Areas (SPAs) under the Birds Directive, which is already in place. These measures aim to maintain and, where necessary, restore the range and variation of the natural habitats and species concentrations for their maintenance on a long-term basis. The selection of SACs takes place at a national level, requiring specialist knowledge of the range and distribution of the habitat types, their size and importance on a European scale.

The Council of Europe funded several reviews of the coastline of Europe in the 1980s. These include reports on saltmarshes (Dijkema, 1984), dunes and shoreline vegetation (Géhu, 1985) and marine benthos (Mitchell, 1987). An overview of such habitat sites is a prerequisite for any identification of sites selected to form part of a Europe wide series of protected coastal areas within the context of the EC Directive on the Conservation of Habitats and Species.

The Science Commission for the European Union for Coastal Conservation is promoting the production of a series of habitat inventories to identify the location and importance of the main coastal habitats. This will provide an overview of the coastal habitats of Europe. Doody (1991) provides an example of this approach.

The Habitats Directive is a very important measure for promoting marine and coastal nature conservation. Its implementation, however, lags behind that of the terrestrial sites (Anon, 1997a). The following summary of guidelines for implementing the Habitats Directive in marine and coastal areas was produced at a seminar, held in Morecambe, organised by the Institute for European Environmental Policy. It introduces three separate stages (Randall & Doody, 1995).

Site identification and management in which it defines a two-stage approach, the identification and selection of the site followed by the development of management strategies:

- best available data on the location, extent and distribution of the habitats and species is needed to enable the establishment of boundaries;
- researching and understanding processes, and the dynamics of the marine environment will enable assessment of site dynamics and sensitivity to human impacts, and future management needs;
- site-specific conservation objectives should be established, followed by an agreement on management measures;
- strategies for achieving conservation objectives should include the use of indicators, monitoring, and evaluation mechanisms to review measures.

Integrated approach, which is often central to the successful management of coastal and marine sites:

- conflicts should be managed and resolved through local co-operation, though resolution through more formal regulations may be required;
- stakeholders are to be involved early on in the management process, and presented with clear proposals;
- management frameworks should embrace the range of relevant interests, ensure practical application of management plans and be adaptable;
- advantage should be taken of existing initiatives.

Supporting measures in the form of policy tools may be useful in establishing the required forms of management and may include the following:

- funding for site identification, designation, management, and developing skills. The possibility of using Structural Funds should be explored;
- practicable monitoring and enforcement arrangements which are clearly tied to the decision-making process. Technological advances

and self-regulation by users may play an important role;
- information and awareness to foster understanding, compliance and support for management measures;
- co-operating with and learning from other projects, including the EU Demonstration Programme on the Integrated Management of Coastal Zones.

The problems currently being experienced of selecting SACs, can be exemplified as follows. The selection can be designated under the Habitats and Species Directive and takes place according to the Criteria set out in Annex 3 of the Directive. This is a two-stage process involving firstly an assessment of the relative importance of each site for the habitat at a national level and, secondly, an assessment of the European (community) level of importance. The specific complications of selecting SACs in the UK are outlined below.

Site selection and conservation in the UK

Taking the example of a detailed analysis and survey of shingle sites in the UK, the range and complexity of the habitat types must be selected to represent the sites of European significance. Key factors in their selection are; size, the presence of geographical elements in the plant communities, the presence of rare plants and the integrity of the structure and function of the single feature. Because of this last consideration, it is possible that several of the sites selected are complex mosaics of habitats and may have been identified for other habitat and species criteria. These might include both the successional vegetation types associated with the formation or a complex of related formations (e.g. North Norfolk where sand dune, saltmarsh, shingle, mud/sand flats and other marine habitats occur together). Other interests may be important, such as their value for important populations of birds or invertebrates, or the presence of rare species.

The existing conservation designations for the UK have provided some protection for vegetated shingle, although this has rarely been the sole rationale for designation. Approximately 200 SSSIs contain shingle though in most cases this takes the form of a fringing beach. Doody (1989) identified 22 of these as containing significant areas of stable or semi-stable shingle vegetation. Several shingle sites form part of NNRs and others are managed by local authorities as LNRs. Of the 10 National Parks in England and Wales, three include coastal shingle. Of the English and Welsh coastline, 40% is designated 'Heritage Coast' and this includes the North Norfolk, Suffolk and Dungeness sites, as well as many of those in Wales. Others, such as Chesil Beach and Slapton Leigh are also AONBs. The North Norfolk coasts and Bridgewater Bay are biosphere reserves. Finally, there are sites such as those owned by non-governmental organisations (NGOs), such as the National Trust and the Royal Society for the Protection of Birds (RSPB).

Despite the wide range of protection measures available in the UK and across Europe, sites still suffer destruction from development. Particular problems include gravel extraction, coastal defence work, military activity, agriculture, forestry and recreational pleasure.

Coastal Zone management in the United Kingdom

The coastline of the UK is an immense and varied resource. Along the 18,000 km of its length, the importance of the coast varies greatly for different uses and interests. Some areas (perhaps because of degradation as a result of historic uses) are of little intrinsic interest, whilst others are of immense commercial value.

Historically, competing demands for space on the coast has resulted in wildlife interests becoming marginalised. Estuaries, in particular, have fared badly with 25% of the intertidal resource lost since Roman times, mostly to agricultural land claim. (Davidson *et al.,* 1991). Significant loss of other coastal habitats has also occurred. For example 14% of sand dunes have been lost in the last 100 years due to afforestation. (Clarke, 1991). Even if losses to land claim are halted in the future, significant areas of coastal habitat will be lost due to rising sea levels unless remedial action is taken. In England alone, as much as 5% of salt marsh and 3% of intertidal flats could be lost (Pye *et al.,*1993).

Management schemes for Marine SACs

Because of the complex nature of the UK's coastline, new legislation has been established to cover the marine environment to meet the terms of the Habitats Directive. This is currently being implemented using 12 demonstration management schemes around the UK and involves a consortium of public and private organisations led by English Nature. The process is being designed to test and improve upon the current best practice in a range of areas, from scientific research to the development of different management approaches.

Within each management scheme, the emphasis will be placed upon partnerships with the relevant authorities and interest groups on the sites. It is intended that the nature conservation agencies will deal with setting conservation objectives for the sites, gather research material, and establish the range of activities that might damage or disturb the key features. A special management group (to be set up individually for each site) will tackle the wider ranging issues affecting the ways in which the site is being used.

The management group, in bringing together interests other than those of nature conservation, will help provide the basis for an assessment on how current site uses might affect the conservation interests. This process of consultation, backed by the best available scientific information, will lead to the development of an agreed management scheme, which will be periodically monitored to ensure that nature conservation objectives are being met. The experiences gained from these demonstration schemes should pave the way for conservation of all marine SACs in the UK and should also help inform similar processes in other Member States (Torlesse, 1997).

Currently, instruments already exist in England and Wales in the form of designated Sites of Special Scientific Interest (SSSI). Similar designations also exist in Scotland and Northern Ireland. The purpose of the SSSI series is to identify those key areas where nature conservation should be the primary objective of site management (DoE, 1994). By 1989, 140 estuaries were notified as SSSIs covering approximately 74% of the national estuarine area. However, SSSIs only cover the coastal sites down to mean low water tide mark (mean low water spring tide mark in Scotland) and is, therefore, unable to address site protection across the whole coastal zone. Those sites, which are of international importance for nature conservation, should be classified as Special Protection Areas (SPAs) under the Birds Directive and/or designated as Special Areas of Conservation (SACs) under the Habitats Directive. All classified SPAs are also notified as SSSIs and therefore suffer under the same limitations (i.e. restricted to terrestrial and inter-tidal habitats). As a result, the Birds Directive, as implemented by the UK Government, can only make a limited contribution towards site protection within the coastal zone. By July 1994, 36% of the estuaries that had qualified, had been classified as SPAs.

With the introduction of legislation directed towards the marine environment (defined as including habitats below the high water mark (HMSO, 1994), SACs should not be restricted to sites first notified as SSSIs and should, therefore, provide for the first time the potential for site protection across the coastal zone. As a result, the Habitats Directive should have an important role to play in site protection within the coastal zone.

Some areas of the coast are of exceptional nature conservation importance. Such sites may be designated as National Nature Reserves (NNNRs), as Marine Nature Reserves (MNRs) or as Areas of Special Protection (ASPs). Collectively, these cover only a relatively small area. Whilst the NNRs are restricted to terrestrial and intertidal habitats, MNRs may include marine habitats.

Planning procedures within the coastal zone

Under the various Town and Country Planning Acts, the planning of building development influences the use of the coastal zone and also the ways in which site protection can be implemented. In the UK, there is a well-developed hierarchical structure to planning. At a national level, the Government produces a series of sectoral Planning Policy Guidance (PPG) Notes. These are all of relevance to the coastal zone. In particular, the PPG on coastal planning attempts to provide guidance for integrated planning within the coastal zone. More recently, the Government has published guidance on nature conservation (PPG9) which addresses the international wildlife obligations, including the Birds and Habitats Directive. This publication identifies the relevant sections in the Statutory Instruments used to implement the Habitats Directive and also identifies what is required of planning authorities by law when reviewing and making planning decisions.

At a regional level, Regional Planning Guidance (RPG) notes are provided. Whilst these are focused on particular regions, there are no formal provision for regional coastal zone planning (although some regional planning conferences have attempted to address this problem (SERPLAN, 1992). At a county level, the guidance provided by RPGs is interpreted within Structural or Unitary Development Plans (where Unitary Authorities have replaced two-tier local government) and from thence into Local Plans at District levels. These plans usually involve the zoning of land use and demonstrate a long history of zoning use within the terrestrial portion of the coastal zone. At a local level, they are one of the principal mechanisms for delivering site protection and resolving conflicts through consultation and enquiries over development plans and individual applications. Development plans are restricted in what they can address. They only deal with physical, built environment issues and are restricted to the Planning Authority, which is normally above the high tide mark and are based upon political rather than natural areas and are, thus, not specific to the coastal zone. Therefore, despite the requirements of the Habitats Directive relating to the protection of marine habitats, there is not an equivalent planning system within the marine environment.

Development Plans have a limited significance for the coast. Firstly, they are linked directly to the administrative areas of the local authorities, which seldom fit the natural areas appropriate for coastal zone management. They normally extend only to the low water mark. Secondly, although reference might be made to wider environmental, social and economic contexts, the plans are limited to land use and planning matters. In recent surveys of coastal local authorities (Taussik, 1993, 1995a), it has been suggested that coastal issues are assuming a much higher priority in development plans than it did in the past. This can, in part, be attributed to the Government guidance on coastal planning, issued in 1992 (DoE & WO, 1992). The scope for policy offered by this guidance is very broad and includes development in general, coastal protection, nature conservation, flooding, ports and harbours, as well as industry, housing, landscape protection and enhancement, population, recreation, leisure and tourism.

One of the greatest limiting factors of development plans relates to the perception of the coast. Planning authorities seldom consider all three of the elements, which comprise the coast; the land, the inter-tidal zone, and the sea. Policy statements normally relate to the land, although a limited number of authorities refer to water quality, views across the water, marine wildlife, the transportation of sea-borne materials, etc., but these are very much the exception. As in most other policy aspects of coastal development plans, policies for wildlife relate principally to land-based wildlife. Many policies make reference to coastal locations or features although few relate to specific habitat types, which has disappointed many conservation bodies who would prefer to witness more specific policies (Dodd & Pritchard, 1993). Although development plans can affect wildlife and habitats in many ways, few plans refer to it *per se*. The policies are normally vague and mainly relate to the inter-tidal zone. Few, however, refer to the impact of land-based developments upon intertidal and marine features, or on hydrological and other processes which affect coastal ecology.

Other hierarchical, sectoral planning systems are beginning to appear although they are non-statutory. A good example is that of the planning of coastal and flood defences. At a national level, the Government's policy towards coastal and flood defence is defined in the Ministry of Agriculture's National Coastal Defence Strategy (MAFF, 1993). This national strategy is translated at regional levels in the form of Shoreline Management Plans (SMPs). These differ from RPGs in that they are not based upon political boundaries but deal with coastal defence issues within coastal process areas called sediment cells. These areas have been recommended by the Environment

Committee (Environment Select Committee, 1992) as the basis for a useful unit in which could be developed an integrated CZM. The function of SMPs is to outline the strategic objectives for coastal and flood defence within a sediment cell. These objectives then guide coastal defence strategies, which are based on a smaller unit of area, such as an estuary.

Whilst the development planning system has provided a mechanism for resolving conflict for many years, The Habitats Directive Statutory Instruments, used in conjunction with PPG9 now provide the a means by which conflicts can be resolved for those developments which do not necessarily fall within development plans. For example, the decision making process outlined in PPG9 may now be applied to the development of coastal and flood defence. The requirements of the Habitats Directive go further than the consideration of individual proposals. All plans and projects, either individually or in combination, which are not directly related to site management but are likely to have a significant effect on the site, should also be assessed in view of the conservation objectives of the site (Article 6[3]). Such an integrated assessment of individual plans within the coastal zone would almost certainly necessitate the production of integrated coastal zone plans. This has not yet happened in the UK.

Management in the coastal zone

Whilst the general approach to planning in the coastal zone is hierarchical albeit on a sectoral basis, the situation regarding management is reversed, with the Government being committed to a bottom up approach. This is based upon individual coastal and estuarine management plans aimed at stimulating the co-ordination of activity management. The greatest problem relating to these management plans is that they are being developed in the absence of regional coastal zone management guidance and in a national policy vacuum.

With respect to the interests of nature conservation, the level of management for any particular area needs to be commensurate with level of importance of the nature conservation, and to the potential for conflict and risk of damage. The same principle would apply equally to management on the grounds of safety. Logically, this results in a hierarchical approach to management. Where there is little or no nature conservation interest, then little or no management control is needed for conservation purposes. Conversely, as nature conservation increases, so does the need for control and sites, which are of exceptional nature conservation importance, have the highest degree of control, which might include the removal of traditional rights such as access

73

and fishing. Concern has also been expressed by coastal users that because the UK coast is of such great nature conservation importance, it will lead to widespread activity bans. This, however, is not an inevitable conclusion since a variety of temporal and spatial variables can be superimposed. For example, the demands on intertidal habitats by birds are seasonal. As increased numbers of birds arrive during winter, the degree of regulation needs to increase but can be relaxed during the summer providing the activities do not result in long term damage to the habitat.

In a similar manner, the management of activities within SACs and SAPs should receive a similar degree of consideration as site development. Sites of international importance cover a smaller area than sites of national importance, have greater conservation interest and require a greater degree of control over the range of permitted activities. The Government approach is currently to deliver coastal management through voluntary management plans. Many such plans cover SPAs and are dependant upon agreements being reached on the objectives of the particular site. However, many of the objectives for SPAs and SACs are already set within European law and the Habitats Directive requires the production of appropriate management plans when needed (Article 6[1]). It might be argued that the necessary conservation measures for SPAs and SACs are in the process of being integrated into development plans. For many coastal SPAs and SACs, however, this will prove to be inadequate because the plans are limited to land based developments and cannot address management issues, either on land or sea.

As SACs are identified and designated, new management mechanisms will be available, particularly for marine sites. The Habitats Directive Statutory Instruments provide for relevant authorities to establish a management scheme under which their functions may be used to secure the conservation objectives of the site. The relevant authorities including the statutory nature conservation agencies, local planning authorities, the National Rivers Authority, harbour authorities and Sea Fisheries Committees amongst others. Since only one scheme may be produced for each site, this will demand an integrated approach. The extent to which such management schemes can or will span the land/sea boundary remains unclear. There must also be some doubts concerning the ability of some relevant authorities to exercise their powers for nature conservation purposes. Thus, the system for CZM within the UK is both complicated and fragmented.

A system, however, must be valued by its results. In England and Wales, about 44% of the coast has protective designations related to landscape quality (National Park, AONB, or Heritage Coast). In addition, substantial lengths of the coast are protected by statutory designations related to nature conservation. The NT and similar bodies are continuously buying more land to be protected.

In contrast to this, Finland has approximately 3% of the coastline under statutory protection (National Park or Nature Reserve) and of this, more than 90% of the protected coast consists of islands. Protection of the mainland coast has become so expensive that neither the municipalities nor the state treasury is willing to pay the money needed for protection. This is also reflected in land use planning. The plans recognise that resources are limited and as a result the coastline is being rapidly colonised by private summerhouses. They can be built almost to the shoreline because there is no tide and the houses and their courtyards prevent access for others. Sustainable use of the coast is, therefore, a difficult objective under these circumstances.

In contrast, the coastal zone of Wales, for example, supports a wealth of natural habitats. Only about 30% of the coastline fails to qualify for some form of protection or landscape designation. The other 70% have a variety of UK designations including statutory protective designations of important biological and geological features and designations designed to preserve some outstanding coastal landscapes. The Welsh coastline also has a number of international designations, including:

- 3 coastal Ramsar sites, designated under the Ramsar Convention on Wetlands of International Importance;
- a coastal Biosphere Reserve (the Dyfi estuary), designated under the UNESCO Biosphere Programme;
- 8 coastal SPAs designated under the EC Directive on the Conservation of Wild Birds.

The marine and coastal areas proposed for possible designation as SACs under the EC Habitats Directive will add another important European tier of protection to this largely undisturbed coastline. Welsh coastal SPAs and SACs could make a significant contribution to the European ecological network, Natura 2000. The proposed coastal SACs in Wales can be roughly divided into sand dunes, sea cliffs, saltmarshes, and marine habitats including estuaries, littoral and sub-littoral sand and mudflats, reefs and a large variety of bays and inlets.

Conclusion

Although there are numerous examples of attempts to integrate management in the coastal zone at all levels, no one approach provides all the answers. This is hardly surprising given the complexity of the coastal zone and the number of human activities, which can have an influence upon it. International agreements have concentrated upon policy issues, largely related to pollution and fisheries, which transcend international boundaries. Examples include the Action Plan for the Mediterranean (Blue Plan, 1993) and the Ministerial Conferences (North Sea Task Force, 1994). These initiatives, however, fail to deal with the land-based problems, such as the protection of species and habitats, or the location of land-based sources of pollution or other issues which cross the coastal/marine boundary. There are very many encouraging moves within Europe to address the issues of regional coastal zone management, although these are very much at the development stage and are dependent upon consistent and reliable data. The problems of CZM must, therefore, be addressed at a national level and this paper has considered many of the problems related to the UK.

Good data and information are vital tools for policy implementation. We need to have an ability to quantify the scale, location and importance of any resource of human value, but we also need to understand how human activities affect that value and can be accommodated in an environmentally sustainable way.

The coastal zone provides a wide range of resources, the exploitation of which results in conflicts both between interest groups and the environment. In most countries, mechanisms exist to manage and plan the different resources. Inevitably, the responsibility for the management of the coastal zone rests largely in the hands of the 'local' communities. In the UK, this is the Town and Country Planning system, which has been traditionally involved in:

- the management of land and property resources;
- resolving conflicts between different interest groups;
- the mitigation of the external effects of development.

Taking, for example, the problems of dealing with coastal zone management within the UK, with the responsibility resting in the hands of the locally elected members of the community, the ultimate goal of seeking partnerships for co-operation throughout Europe might appear to be unrealistic. However, there are sufficient existing frameworks throughout Europe to suggest that a Pan European, Integrated Coastal Zone Management policy is a realistic aim. Such a policy could include investment strategies, integrated development plans and resource management strategies. There must, however, be a common strategic understanding of how European coastal areas and resources can be organised to meet future needs in a sustainable manner. This can only come from a formal agreement among the states of Europe to improve the planning and management of human activities within the coastal zone, within the context of the existing environmental laws, through the implementation of integrated coastal zone management. This strategy must reflect the needs of the states together with the challenges of sustaining the Natura 2000 network.

References

Alexander, L.M., 1990. The co-operative approach to ocean affairs: twenty years later. *Ocean Development and International Law*, **21**, 105-109.

Amselek, P., Cohen, J. & Prieur, P., 1994. *Legislative Measures Taken or to be Taken by the Member States of the Council of Europe for the Protection of the Coastline*. European Information Centre for Nature Conservation, Council of Europe, Strasbourg.

Anon, 1996a. Commissioner proposes measures to halt further decline in the EU's wetlands. *Natura*, **1**, 8-9,
http://europa.eu.int/en/comm/dg11/news/natura/nat1en.htm (07/04/98).

Anon, 1997a. Implementing the Habitats Directive in marine and coastal areas. *Natura*, **4**, 2-5, http://europa.eu.int/en/comm/dg11/news/natura/nat4en.htm (07/04/98).

Blue Plan, 1993. *Overview of the Mediterranean Basin (Development and Environment)*. Mediterranean Action Plan (1st Phase), Marseilles.

Carter, R.W.G., 1988. *Coastal Environments: An Introduction to the Physical, Ecological and Cultural Systems of Coastlines*. Academic Press. London.

Carter, R.W.G., 1989. Coastal zone management: comparisons and conflicts, Proceedings of the Symposium of *Planning and Management of the Coastal Heritage*, 45-49, Southport.

Council of European Communities, 1992. *Resolution 92/c59/02 of 1st February on the Future Community Policy Concerning the European Coastal Zone*. No. C59/1, 06/03/92.

Council of European Communities, 1995. *Progress Report on the Implementation of the European Community Programme of Policy and Action in Relation to the Environment and Sustainable Development Towards Sustainability*. COM(95) 624 Final.

Council of European Communities, 1996. *Demonstration Programme on Integrated Management of Coastal Zones*. European Commission Services, Information Document XI/79/96.

Countryside Commission, 1970. *The Planning of the Coastline. A Report on a Study of Coastal Preservation and Development in England and Wales*. HMSO, London.

Department of the Environment & The Welsh Office, 1992. *Town and County Planning (Assessment of Environmental Effects) Regulations*. Circular 15/88, HMSO.

Department of the Environment, 1993. *Managing the Coast*. HMSO. London.

Department of the Environment, 1994. *Planning Policy Guidance: Nature Conservation (PPG9)*. HMSO, London.

Dijkema, K.S., (ed.) 1984. *Salt Marshes in Europe*. European Committee for the Conservation of Nature and Natural Resources. Council of Europe, Strasbourg.

Dixon-Gough, R.W., Mansberger, R. & Seher, W., 1998. Global, regional and local policies for land conservation and planning. *27th International Symposium of the European Faculty of Land Use and Development*, Zurich.

Dobbin, J.A., 1993. Environmental programme for the Mediterranean (EPM) and municipal environmental audits. *EMECS '93, Environmental Management of Enclosed Coastal Seas: Abstracts*, 83.

Doody, J.P., 1989. Conservation and development of coastal dunes in Great Britain. In: Van Meulen, F., Jungerius, P.D. & Visser, J.H. (ed.), *Perspectives in Coastal Dune management*, SPB Academic Publishing, The Hague.

Doody, J.P., 1991. *Sand Dune Inventory of Europe*. Joint Nature Conservation Committee, Peterborough & EUCC, Leiden.

Doody, J.P., 1995. Infrastructure development and other human influences on the coastline of Europe. In: Salmon *et al.* (ed.), *Proceedings of the 4th EUCC Conference Marathon, Greece, 1993*.

EC, 1999a. *Towards a European Integrated Coastal Zone Management (ICZM) Strategy. General Principles and Policy Options*. Directorates-General Environment, Nuclear safety and Civil Protection, Fisheries, Regional Policies and Cohesion, European Commission, Office for Official Publications of the European Communities, L-2985 Luxenbourg.

EC, 1999b. *Lessons from the European Commission's Demonstration Programme on Integrated Coastal Zone Management (ICZM)*. Directorates-General Environment, Nuclear safety and Civil Protection, Fisheries, Regional Policies and Cohesion, European Commission, Office for Official Publications of the European Communities, L-2985 Luxenbourg.

Environment Select Committee, 1992. *Coastal Zone Protection and Planning*. House of Commons Environment Select Committee Second Report, Session 19991-92, HMSO, London.

Géhu, J-M., 1985. *European Dune and Shoreline Vegetation.* Nature and Environment Series, (32), Council of Europe, Strasbourg.

Gubbay, S., 1994. *Coastal Zone Management and the North Sea Ministerial Conference.* WWF/MCS Report.

Haas, P.M., 1990. *Saving the Mediterranean.* Columbia University Press, New York.

Haas, P.M., 1991. Save the seas: UNEP's Regional Seas programme and co-ordination of regional pollution control efforts. In: E.M. Borgese & N. Ginsberg, (ed.), *Ocean Yearbook 9.*

Healy, M.G., 1995. European Coastal Management: An Introduction. In: Healy & Doody (eds.), 1995. *Directions in European Coastal Management,* 1-6, Samara Publishing Ltd., Cardigan.

Hinrichsen, D., 1990. *Our Common Seas: Coasts in Crisis.* Earthscan Publications, London.

HMSO, 1994. *Planning Policy Guidance: Nature Conservation (PPG9).* HMSO, London.

Huggett, D., 1995. The role of the Bird's Directive and the Habitat and Species Directive in delivering integrated coastal zone planning and management. In: Healy, M.G. and Doody, J.P., (eds.), *Directions in European Coastal Management,* 8-18, Samara Publishing Ltd., Cardigan, UK.

Huggett, D., 1996. Progressing coastal zone management in Europe: a case for continental coastal zone planning and management. In: Taussik, J. & Mitchell, J. (eds.), *Partnership in Coastal Zone Management.* Samara Publishing Ltd., Cardigan, UK.

Hughes, T.P., 1994. Catastrophes, phase shifts and large-scale degradation of the Caribbean coral reef, *Science,* **265,** 1547-1551.

Hulm, P.M., 1983. The Regional Seas Program: what fate for UNEP's crown jewels? *Ambio,* **12**(1), 2-13.

Jacobson, M.A., 1995. The United Nations' Regional Seas Programme: how does it measure up? Coastal Management, **23,** 19-39.

Ketchum B.H. (ed.), 1972. *The Waters Edge: Critical Problems of the Coastal Zone.* MIT Press, Cambridge, Ma., USA.

Land Use Consultants, 1990. *Countryside and Nature Conservation Issues in District Local Plans.* Countryside Commission, Cheltenham.

MAFF, 1993. *Strategy for Flood and Coastal Defence in England and Wales.* Ministry of Agriculture, Fisheries and Food / Welsh Office.

Miles, E.L., 1983. On the role of international organisations in the new ocean regime. In: Park, C-H., (ed.), *The Law of the Sea in the 1980s.* 383-445, University of Hawaii Press, Honolulu.

Mitchell, R., 1987. *Conservation of Marine Benthic Bionoses in the North Sea and the Baltic.* Council of Europe, Strasbourg.

Nature Conservancy Council, 1984. *Nature Conservation in Great Britain.* NCC, Peterborough, UK.

North Sea Task Force, 1993. *North Sea Quality Status Report.* Oslo and Paris Commissions.

Pye, K. & French, P.W., 1993. *Targets for Coastal Habitat Creation.* English Nature Research Report 35, English Nature, Peterborough, UK.

Randall, R.E. & Doody, J.P., 1995. Habitat inventories and the European Habitats Directive: the example of shingle beaches. In: Healy, M.G. & Doody, J.P., (eds.). *Directions in European Coastal Management,* 8-18, Samara Publishing Ltd., Cardigan, UK.

Ruddle K., Morgan W.B. & Pfafflin J.R. (eds.) 1988. *The Coastal Zone: Man's Response to Change.* Harwood Academic Press, Geneva, Switzerland.

Sand, P.H., 1988. *Marine Environmental Law in the United Nations Environment Program: an Emergent Eco-Regime.* Tycooly Publishing, London.

SERPLAN, 1992. *Coastal Planning Guidelines for the South East.* A report of the Coastal Policies Working Party.

da Silva, M.C., 1996. Coastal zone management initiatives of the European Environment Agency. In: Taussik, J. & Mitchell, J. (eds.), *Partnership in Coastal Zone Management.* Samara Publishing Ltd., Cardigan, UK.

Siren, 1992. Regional news. *The Siren: News from UNEP's Oceans and Coastal Areas Programme.* **46**(June), 32.

Siren, 1993. Reviewing MEDPOL. *The Siren: News from UNEP's Oceans and Coastal Areas Programme.* **49**(Oct/Dec), 21-24.

Taussik, J., 1993. *Development Plans: Effective Involvement in Plan Preparation.* Working Paper, Department of Land and Construction Management, University of Portsmouth.

Taussik, J., 1995a. *Development Plans in the Coastal Zone.* Working Paper in Coastal Management (13), University of Portsmouth.

Taussik, J., 1995b. The contribution of Development Plans to coastal policy. In: Healy, M.G. & Doody, J.P., (eds.), *Directions in European Coastal Management,* Samara, Cardigan, UK.

Tooley, M.J., 1987. Sea level studies. In: Tooley & Shennan (eds.) *Sea Level Changes,* 1-24. Basil Blackwell, Oxford.

Torlesse, J., 1997. The UK experience: management schemes for marine SACs co-financed by LIFE (Nature). *Natura,* **4**, 4, http://europa.eu.int/en/comm/dg11/news/natura/nat4en.htm (07/04/98).

UNEP, 1982a. *Achievements and Planned Development OF UNEP's Oceans Regional Seas Programme and Comparable Programmes Sponsored by other Bodies.* UNEP, Nairobi, Kenya.

UNEP, 1982b. *Convention for the Protection of the Mediterranean Sea Against Pollution and Other Related Protocols.* United Nations, New York.

UNEP, 1991. *Status of the Regional Agreements Negotiated in the Framework of the Regional Seas Programme.* UNEP, Nairobi, Kenya.

UNEP, 1993. *Catalogue of Publications, No. 10.* United Nations Environmental Programme. Nairobi, Kenya.

Vallega, A., 1992. From the action plan to the Mediterranean Agenda 21. In: Ozhan, E. (ed.), *MEDCOAST 93: Proceedings of the First International Conference on the Mediterranean Coastal Environment,* 1-12.

5 Interdisciplinary approaches in coastal zone management and floodplain area development

U. LENK AND H. LENK

Introduction

General introduction to interdisciplinarity

Already in 1970, one of the present authors stated (Lenk, 1970) that an information and systems technological age and society would be developing and increasingly gain-profile and impact. Indeed, there is a rather encompassing trend towards cross-disciplinary systems in a progressingly interlaced world as it may be discovered or to a considerable part made or encroached upon by man. This development comprises trends which have since then dramatically accelerated to get weight and prominent position in the last two decades and are still progressively getting more and more comprehensive impact to manipulate and reshape if not revolutionise our environment and the social world. We seem to live in a rather socio-technological, a manmade technological and thus in a sense an 'artificial' world - at least to a considerable and increasing degree.

Systematic methods and methodologies prevail. Essential are not only methods, but also methodologies: this trend is to be found in all science-induced technological developments as well as in administrations, general trends characterising ever growing fields to be captured by operations technologies led by (methodical or even methodological) process controlling and systems engineering, by operations research etc.

Informatisation, abstraction, formalisation and concentration on the operational are essential. It is by the way of computerisation and informatisation as well as by using of formal and functional operations technologies (e.g. flow charts, network approaches etc.) that the formal essentials of increasingly comprehensive processes, organisations as well as the interrelations of different fields and sub-fields are integrated. Information technologies lead the way.

For comprehensive systems engineering or system technology, it is indeed characteristic that the different technological developments including economic and industrial realms are getting a joint impact providing the development with a kind of positive feedback leading to system(at)ic interaction and generally to a kind of systems acceleration across different fields. (This is a trend which had been predicted by Gottl-Ottlilienfeld (1923) already in the nineteen-twenties: 'mutation', 'filiations' and intensification of different processes across traditional realms: mutually interactive spillover effects and a sort of what we nowadays would call positive feed-back processes.)

All these ongoing processes necessarily require a far-reaching, if not encompassing interdisciplinary interaction and stimulation ('interstimulation'). Indeed, interdisciplinarity led by spillovers from science to science and from thence to technological development and innovation (as implementation) as well as to society at large would characterise the embedding of the interdisciplinary interactions into the overall developmental purview. Systems analyses and systems technologies require interdisciplinary approaches in practice. The pertinent challenges within this 'world of systems' including technosystems, social systems and ecosystems requires a thorough methodological study for the respective approaches and types of interdisciplinarity in research and development (R&D).

Short of providing such a methodological analysis here, it may suffice just to mention criteria for the methodological distinction of disciplines according to:

- objects and areas as well as scopes;
- (arsenals of) methods;
- knowledge interests (Habermas, 1968);
- theories and their systematic and historical connections and contexts (points 1 through 4 after Krüger, 1987);
- relationship between theory and practice;
- substantivity versus operationality of theories (substantive versus operative theories) (Bunge, 1967);
- systems holisma versus realm specifity;
- *a priori* analysis and formal methods versus empirical analyses;
- patterns of explanation and systematisation (descriptive versus explanatory, historical versus systematical);
- cognitivity and normativity (descriptive versus normative disciplines);

83

- fictive, virtual (secondary) or social validity versus primary reality.

These divers and diverse views regarding the distinction of disciplines certainly lead to different types of bi- and multilateral interdisciplinary relationships between the respective disciplines.

Types of interdisciplinarity

The following levels of interdisciplinarity are considered, below:

- interdisciplinary co-operation in more or less well-defined projects;
- bi-disciplinary or interdisciplinary research area;
- multi-disciplinary aggregate field of research (e.g. environmental research);
- genuine 'interdiscipline' (like physical chemistry or biochemistry);
- multi-discipline resulting from/relying on multidisciplinary theoretical integration;
- abstract generalised interdisciplinary systems theories (e.g. general systems theory);
- mathematical theories of abstract and complex dynamical systems (e.g. deterministic or a less developed probabilistic chaos theory);
- supra-disciplinary abstract structure-analytic and operational disciplines (e.g. operations research);
- methodological supra-discipline as e.g. philosophy of science and science of science;
- philosophical and methodological epistemology as a meta-disciplinary approach (for example, the methodological schema interpretationism, cf. Lenk, 1993;1995a,b).

At first, we have just the simple co-operation of different experts for, or within, a developmental programme such as in coastal zone management planning. Here experts from different fields such as geography, cartography, hydrography, geodesy, biology and ecology, limnology or oceanography, as well as engineering in dike-building and landscape planning have to co-operate. Secondly, an interdisciplinary or bi-disciplinary realm of research might evolve or, thirdly, as a multidisciplinary aggregative research area. The fourth level or step of co-operative integration would amount to a real interdiscipline (like molecular biology) or, fifthly, a multidiscipline in the more specific sense (multidisciplinary theoretical integration). The sixth through eighth levels are formal theories of an abstract mathematical brand being used as instrumental vehicles of modelling real or constructed systems. Furthermore, the meta-theoretical levels 9 and 10 are addressed to a higher stage of methodological or epistemological (meta-) analyses.

Interesting questions regarding geographical information systems (GIS) and their application to coastal zone management are:

- on what level are the actual and potential interdisciplinary co-operations in both of these fields to be located?
- how can possible and already existing levels and types of interdisciplinary interact with one another?
- is it possible to distinguish and effectively separate descriptive and normative utilisation of interdisciplinary modelling, e.g. with respect to factual ('cognitive') and interest or value conflicts (see below)?
- can we neatly distinguish between scientific and purely descriptive valuative approaches in the practice of system planning, to wit e. g. coastal or shore zone management?
- to what extent are value orientation and interests ('humanware' so to say) indispensably moderating variables for any application of GIS and planning procedures, e.g. coastal management acts and plans (see below)?

With respect to the evolved types or stages of interdisciplinarity, we would hypothesise and argue that the practical elaboration of GIS and the interdisciplinary collaboration in landscape, land-use and coastal as well as lake and river shore management have, thus far, not progressed beyond level 3 and possibly only as far as level 2. In the foreseeable future it is hardly likely to reach level 5 of a really theoretical multidisciplinary integration. Yet, advancing interdisciplinary approaches in all of these mentioned fields will turn out to be necessary for and conducive to practical applications in the near future - so at least we will argue in the main part of the paper addressing the feasibility of GIS to coastal management and planning problems.

Introduction to the topic

Increasing awareness of the decline of nature resources renders it necessary to re-think the traditional approaches to nature resource management. Governments all over the world are setting up programmes to establish coastal zone management for their shores (e.g. NOAA, 1997).

Thus far, most of the time the approaches to land management and coastal zone management are sectoral suffering from the differing uni-disciplinary views of the disciplines involved in planning and management projects in these areas. In fact, sometimes the results of planning processes are not optimised or harmonised with the goals of other disciplines of concern in the areas at hand. For instance, traditional flood defence is conducted on the basis of presupposing that human beings and their property have to be protected from natural hazards, and extensive agricultural use of the area has to be facilitated. Most of the times the development of an ecologically sustainable environment management which is nowadays the aim of landscape planners and biologist is neglected.

Coastal zones and similarly inshore areas such as deltas, floodplain areas or greater lake districts are used in very different kind of ways, such as:

- recreation;
- transport and shipping and their infrastructure;
- economic use such as mining, fishery and aquaculture, energy;
- strategic use;
- dumping site.

Furthermore, these areas may be of ecological importance, may be qualified as a nature preserve and natural resource with, e.g., unique or rare species of coastal life, or even archaeological sites may be there. Hence, all the disciplines involved in the above mentioned activities try to meet their goals in planning processes for these areas.

It has been stated at many occasions and in many publications that there is a requirement for an integrated approach to coastal zone management (e.g. Gubbay, 1990) and, in addition, it is generally accepted that the introduction of information technology to this problem is essential (e.g. Munro & Bradbury, 1996; Stough & Whittington, 1985; Lenk *et al.*, 1997). However, the problem of how to implement such an integrated GIS must still be resolved. Indeed, there are many aspects and problems confronted in interdisciplinary information technology and GIS projects.

The problems may be divided into sections of:

- problems with data, its availability and data modelling including the varying procedures of data processing and evaluation;
- problems concerning the software to be used, i.e. information technology, especially geo-information technology;
- 'humanware' including organisational and institutional problems, especially problems of value differences and interest conflicts.

These problems will be dealt with briefly in the subsequent sections followed by an example of an integrated approach to floodplain area development, which is currently undertaken by three departments of Hannover University, Germany. Finally, the concept of a GIS for the coastal zone will be introduced.

Data modelling in interdisciplinary GIS projects

When an integrated GIS project is planned, the question must be considered which existing digital data sources are available, which one of these should be used and which analogue data would still have to be integrated into the GIS. There are several starting situations possible when planning an integrated GIS:

- *situation A*: all data to be integrated is available in digital format;
- *situation B*: basic digital geo-data sets are available for the area and only user or discipline specific analogue data has to be included;
- *situation C*: only analogue data of the area to be covered is available.

After a rather general introduction into data integration, these situations will be analysed in this paper.

General aspects of data integration or data fusion

Different data models are often related to the format of the data sets at hand and to the software used for the specific GIS projects. The first differentiation is given by the distinction between raster-based systems, where the respective land parcels are described by pixels and vector-based systems, where features and boundaries between features or objects are represented by means of vectors. Vector based systems may be further divided into layer-based systems where the features classes are separated in layers, and object-structured systems where features are encoded by feature class codes and distinguished further by attributes. Some hybrid systems do exist, which are capable of processing both types of data. However, most of the time real hybrid processing of both types of data and switching between them via vector-to-raster-conversion and vice versa is not available yet. In object-oriented GIS either raster-objects or vector objects may be integrated as objects (Woodsford, 1995).

Prerequisites for successful data integration are given in Grünreich (1992b):

- all metric information must be related to a common geodetic reference frame;
- physically identical geometry may only exist once in the data set which is used for analysis, especially the geometry of discrete objects; they are related to topographical objects or other selected objects, e.g. administrative objects;
- in its semantic modelling, the user-specific description of the environment has to establish a relation to the topographic model if features on the earth's surface are being used as their basis; in other cases other thematic features should be used.

When merging vector data from different data sources, one has to cope (particularly in coastal regions) with problems such as:

1. varying datums between data sets:

   - i.e., horizontal datum problems; data of different sources are often related to different geodetic datums, i.e., they have different types of horizontal co-ordinates such as geographical co-ordinates (e.g. on charts) and co-ordinates referenced to a national grid (e.g. on topographical maps); hence, co-ordinates must be converted and appropriate transformation parameters must be available;
   - vertical datum problems between existing data sets, e.g., there is the datum shift between the vertical datum used on land and the vertical datum used in nautical applications, i.e. charts. The latter are normally referenced to a nautical datum such as the lowest

astronomical tide taking into consideration tidal effects of the moon and the sun, the lowest possible low water allowing for meteorological effects, or simply mean low water springs. These datums also vary from state to state (Ingham & Abbott, 1992). Land vertical datums are often referred to the geoid which may be approximated by mean sea level (Torge, 1991). However, these vertical datums may also vary between nations.

2. geometric and topological problems:

   * sliver polygons, over-shoots, under-shoots and gaps, originating when vectors describing the same physical boundary are digitised at different occasions or on the basis of maps of different scales and projections.

3. semantic problems:

   * redundant information in the data sets to be used, e.g. the same real world objects are represented by features types in the respective different data sets; additionally, these feature types may have differing definitions in the respective feature catalogues.

4. problems with different data formats:

   * simple conversion problems between the various formats used in geographical information technology, since it is common knowledge that almost no conversion-filter works without some difficulties occurring.

Datum problems are particularly encountered when data of different states is to be merged, i.e. when setting up a cross-boundary GIS. As the international relevance of coastal zone management increases, the best choice would then be to use international datums. For example, in Europe, the European Reference Frame could be adopted as the horizontal datum and the Unified European Levelling Network as the vertical datum for an international approach in GIS for coastal zone management, providing transformation parameters exist (Grünreich, 1995).

The problems encountered when merging raster data are less difficult as they may normally be solved by applying image processing functions, such as geo-coding and resampling.

Situation A: all data is available in digital format

A view to the GIS market shows that there are several information systems and geographical databases under development. For example, National or Regional Geological Surveys are setting up data bases to describe the soil and its characteristics, and surveying and mapping agencies establish data bases for cadastral, mapping and topographical purposes. An example for the latter may be the Authoritative Topographic Kartographic Information System (ATKIS) in use in Germany (Grünreich, 1992a; OII, 1997). Hydrographic departments all over the world co-operate to set up the Electronic Chart Display and Information System ECDIS (Scheuermann, 1997).

Existing geodata sets may be distinguished with regards to their spatial relationship:

- vertically separated data sets; such as geological and topographical GIS;
- horizontally separated data sets (including overlapping areas); such as ATKIS and ECDIS or data sets of topographical GIS belonging to different states;
- parallel data sets; such as ATKIS and data sets for car navigation such as the Geographic Data File (ECDGXIII, 1997); this may also include data sets differentiated by their scales; such as topographic data sets and real estate information systems.

Vertically separated data sets will rarely have a physically identical geometric framework. For example the boundaries between different types of soil are generally not related to boundaries between topographical objects. Even the semantic information will seldom be redundant in the feature catalogues. Thus, merging data from these types of basic geodata sets will not run into severe geometric and semantic problems. The problems encountered will mostly be due to different data models and formats.

In horizontally separated data sets it turns out to be most likely that there are at least a few redundancies in the feature catalogue, especially in case of overlapping areas. Thus, for example, the ECDIS feature catalogue incorporates also topographical information for the coastal zone which may be important for navigation in the area (IHB, 1993). If data sets of different topographical GIS are to be processed it is obvious that there will be redundancies which have to be harmonised. Additionally, there will normally be difficulties encountered in the utilisation of different data models if such data sets are to be processed. An interesting information about the availability of digital geographical data in Europe is provided by the GDDD-Project (Geographical Data Description Directory) as part of the MEGRIN-Alliance (Multi European Ground Related Information Network, (MEGRIN, 1997)).

Parallel data sets will most likely be confronted with geometric and topological problems as they will have to rely on data models specific to their tasks. A simple example may illustrate that in one data set, e.g. ATKIS, a road may be represented by a single line whereas in another data set, e.g. GDF, the same road is divided further into various lanes (Walter, 1997). As mentioned above, there is also the distinction given by the scales of the data sets to be merged. If these data sets are digitised on the basis of maps of different scales or maintained at different scales, problems concerning the topology and accuracy of neighbourhood will arise.

However, the availability of digital data sources is a great advantage because digitising all data from scratch is fairly cost-intensive. In the last analysis, it seems not to be completely feasible with regard to human interest or value considerations (see below). It is generally advisable to investigate beforehand whether existing data sources may be used. A difficulty that still has to be solved is the problem of how to update a data base if data from a basic geodata base has been taken and altered accordingly for the specific purpose, and the latter data base is updated, i.e. its data sets change in time.

Situation B: only basic geo-data sets are available

A major advantage of the availability of basic geo-data sets (Grünreich, 1992a) or shared land base (Huxhold & Levinson, 1995) is that costs for data collection for the project may be reduced. As a basic geodata set is being used, it is advantageous to use GIS software, which was already corroborated to work properly with its data model. Otherwise, the user would need yet to assure that the software is capable of handling the data model of the basic geodata set. In addition, it would seem necessary to confirm that the software provides the opportunity of extending the object definitions for the new features to be stored. However, this does not guarantee that the system will be able to process the user-specific data. An example is given in Rimscha (1996) where an integrated GIS was planned for the purpose of harbour management for the port of Hamburg. Interestingly, a Europe-wide tender failed in the first instance to locate a company that could satisfy all its requirements. These included the ability to identify a local support office and the ability to handle the user-specific data, i.e. hydrographic data, and basic geodata from the town's survey department and the survey departments of the surrounding countries.

There are three distinct approaches for data integration on the basis of existing geo-data sets. The latter may be used as general background information, e.g. for visualisation purposes, as the common denominator for geometry, i.e. geometrical data integration, and as a basis for geometrical and semantic data integration.

The reasons for geometrical data integration are to avoid the above mentioned geometrical problems such as sliver polygons, gaps in data, over-shoots and under-shoots which might cause troubles of referencing and integrating or of compatibility or might even impair the results during the pertinent data analyses.

The general approach for semantic data integration on the basis of an existing geodata set is that a feature catalogue for the user-specific data is developed and investigated with respect to correspondences with the feature catalogue of the basic geo-data set. There seem to emerge several possibilities as results of the investigation of the features (MUV-B-W, 1996):

- an already defined feature can be used without alteration;
- an already defined feature can be used by changing the feature type, e.g. a user-specific feature has to be represented by an area while it is defined as a point so far;
- the capture definition for a specific feature has to be altered;
- further attribute values for existing attribute types have to be added, i.e., the domain of attribute values has to be extended;
- further attribute types for existing features or objects have to be defined;
- or further features have to be defined to fulfil the requirements of the integrated data catalogue.

The second and third opportunity generally require that features have to be altered with respect to their geometry, i.e., new geometries will be digitised and incorporated into existing features, or existing features will be generalised in order to simplify their geometric shape.

The use of a basic geodata set provides the opportunity to customise existing data capture software in accordance to the list of feature that still have to be digitised. In addition, an application must be created to alter already existing features with respect to the above-mentioned possibilities, i.e., adding new attributes, changing attribute values, creating new features on the basis of an existing geometry, or changing the geometry of some features.

If there are some constraints on the data set defined in the feature catalogue that may be violated by alterations, such as that features of a specific type have to cover the whole project area, or some features are not allowed to overlap, it should be substantiated whether these constraints are further useful. In case of the latter, some checks with regards to these constraints must be provided either by the GIS-software or on the basis of procedures to be implemented. As for further requirements of research, one of the topics as listed above must be mentioned regarding the problem of how to update data, which has been collected from basic geodata sets and changed for the specific task in hand.

Situation C: only analogue data is available

There are several advantages and disadvantages within this situation. The first advantage is that one is not restricted to a certain data model when planning the GIS. All the different types of aspects and views to the real world may be taken into account when the system is designed. The only restriction will be given by the requirement that the software to be chosen must be able to process and store the data model. However, everything else concerning the set-up of the system will be difficult as it is most likely that no existing user-friendly software for data capture may exist.

Existing data sets and their feature catalogues may serve as a guide for the planning of the data model. If the survey departments of the respective state are already busy with the set-up of a basic geodata set, a co-operation with the respective departments is advisable to share the costs of data capture. Similar strategies apply if other departments or companies are in the course of digitising geographical data.

Software

Generally speaking many users of various software products have experienced the following software problems:

- no software is delivered without bugs, as a consequence it is advisable to negotiate that updates and patches for the software will be supplied by the software company on a regular basis;

- GIS software packages become bigger and bigger (unfortunately sometimes even more complicated) and sometimes can no longer be handled by a single person as a system administrator; hence, there is generally the need to have proper customer support by the software company, at least via hotline;

- general operating system problems: there is a strong need for proper operating system administration, since the same reasons apply as for GIS software administration, i.e., growing operating systems which are sometimes not really user-friendly anymore.

Humanware

Value and interest issues

Apart from factual and data aspects, interest and value aspects tie in with any planning project whatsoever. This is true also for the conception of planning in coastal zone manegement - especially for setting up priorities, solving factual description differences ('cognitive') and 'interest (value) conflicts' (Wang & Stough, 1986, as cited by Obermeyer & Pinto, 1994). Only the former type of conflicting factual information selection, (re)construction and/or retrieval may be solved by scientific GIS procedures (Lenk, 1970), real interest discrepancies and value conflicts may not be susceptible to pure scientific methods but would be dependent on the practical establishment of social consent and, usually, the finding of fair compromises among different value orientations or interests, respectively. It might be possible to find a fair compartmentalisation or segregation strategy, e.g. a pie-cutting or spatial or temporal limitation of usage as discussed with respect to the so-called Naturalists' or Enjoyers' Dilemma, a special modification of the Prisoners' Dilemma in economic game theory (Lenk & Maring, 1990). There is a conflict of interests and incompatible usages (notably at one and the same time) between, e.g., anglers and water skiers on a lake. Contracting for segregation and/or spatial or temporal compartmentalisation might provide a helpful strategy of conflict resolution or of conflict mitigation or at least limitation.

In general, however, GIS data and procedures - while being instrumental in spotting and resolving factual (cognitive) conflicts in gathering, selecting and structuring the pertinent data - will not be able *per se* to solve interest and value conflicts. On the contrary, they might only pinpoint rather than render explicit or even exacerbate such conflicts. Obermeyer & Pinto (1994) discuss an interesting example of an interest coalition despite differing values amongst environmentalists and potentially affected home-owners and unionists in the case of the planned siting of a nuclear power plant, the Bailly plant adjacent to the Indiana Dunes National Lakeshore. In this instance, the opponents fought a successful, 10 years legal war utilising 'cognitive conflicts' to support their respective interests.

Values and interests are typical normative interpretative constructs guiding and controlling action and decision making. Value and interest concepts might also be used in a rather descriptive ('secondary') manner by way of explaining, reconstructing or interpreting orientation and decision making from an observer's point of view. In any case, they are typical (normative or descriptive) interpretative constructs mediating between individual and social perspectives (Lenk, 1987). As interpretative constructs, they allow for describing and explaining the role of value judgements and value plus interest orientation in day-to-day actions as well as for delineating justifications of an individual (personally rule-governed) and of a societal (norm-guided) kind. (Social values are interpersonally and culturally engendered or evolved interpretative constructs, which are backed by institutionalised, at times negatively sanctioned norms being internalised via institutionalised behaviour and action expectations. They are able effectively to rule and control actions and action orientation. Moreover, they might figure as indispensable moderator variables in setting up decision making and planning processes. Humanware is therefore indispensable. It has to be taken into account in any process of planning and systematising actions and establishing as well as filling out the necessary frameworks, e.g. of GIS, too.

Organisational issues

Systems engineering is usually by its very nature embedded in broader societal and even political contexts requiring specific organisational co-ordination, implementation planning and management in practice (see Huxhold & Levinsohn, 1995; Campbell & Masser, 1995).

A clear definition of the goals and objectives plus the scope of the systems development and management is especially helpful if different organisations, agencies and experts from very diverse fields are to co-operate in a truly interdisciplinary manner. Implementation here as a rule crosses established organisational boundaries creating additional technical and management challenges. It seems generally advisable that the overall administration of an interdisciplinary project is conducted by a third and maybe fourth party controlling and/or supervising. Alternatively, this could be conducted by a special management or supervisory group, which is in a sense, rather independent from technical matters and would only or at least mainly be concerned with the organisation or evaluation of the project.

Some 'lead agency' of a committee structure should exist from the beginning. It should attempt to work out an organisational plan and planning framework, secure reliability of communication, foster motivation and allocate responsibilities and tasks plus time schedules, identify sources of potential conflicts and failures in advance and specify the co-ordination of the project(s). This is all the more important, if not necessary, in rather complex joint venture projects of different and multi-participating agencies, organisations, or departments working together more or less autonomously. With respect to the project-internal co-ordination, leadership and motivation Huxhold & Levinsohn (1995) interestingly enough coined the term 'interpreneurship', which already points to social psychological phenomena of motivation, leadership and group or team dynamics (see below).

Staff issues

In multi-participant projects of a truly interdisciplinary character not only different organisational agencies and institutions have to co-operate and to co-ordinate their goals, aims, and strategies as well as activities, but divers experts of at times very different provenance and qualification have to communicate and work with one another. Since there is, to date, no common methodological ground of a general and special methodology and philosophy of technology, especially systems theory and system technology available on which to build of, frequently many difficulties of communication and co-ordination between different perspectives and technical as well as practical language routines and traditions are confronted and to be solved. This is true for applied GIS projects of an interdisciplinary kind, too. It would seem recommendable already in graduate programmes to prepare and school the students for interdisciplinary communication, co-operation, and co-ordination. This would require a certain amount of teaching, training and practice in interdisciplinary communication. A general programme in the philosophy and methodology of technology, systems theory and science (philosophy of science) would be as helpful as a common qualification in general and applied computer science and information technology. Only the latter factor seems to be provided in all engineering programmes by now. This is better than nothing, but there still is a real lack of necessary preparative qualifications in the non(-purely)-technological fields like philosophical methodology, social science (and practice!) aspects - including engineering ethics plus management and organisation theories (and practice!) and legal aspects as well. Earlier questionnaires among German engineers - particularly those working in applied engineering - in the seventies already postulated up to 30% of non-technological additional qualifications in methodology, social science, law, ecology, cybernetics and general systems theory etc. (Hillmer *et al.*, 1976). This requirement certainly applies to preparative qualifications for interdisciplinary GIS projects, too. Just computer science and practice is not enough. Since this requirement is not usually met by the participants of a truly interdisciplinary GIS or ecological (coastal zone management and development) project, much time and energy is frequently squandered by building up a communication and common technical, as well as methodological and linguisitic, denominators in time-consuming processes of training on the job with many trials and errors to go through. Training on the job is necessary and conducive though, but it should not start from nothing or just struggle with elementary difficulties to begin with.

Motivation is another important topic for personnel recruitment. If a team takes up a very ambitious or difficult project, similar phenomena and motivational processes occur as in high performance sport teams. Achievement motivation is an important moderator variable as in any joint and group venture whatsoever (Lenk, 1983). Selection of personnel is a tricky and risky enterprise indeed. We cannot here delve into these problems of applied psychology and team dynamics (Lenk, 1977).

Example for an integrated approach to floodplain area development

This section describes an interdisciplinary research project which is currently undertaken by three institutes at Hannover University, Germany, and follows mainly Lenk *et al.* (1997). The project at hand is concerned with a new interdisciplinary approach to floodplain area development and how GIS, basic geodata sets and cartographic visualisation can support the latter.

Similarly as in coastal regions, the conflict in floodplain area development evolves from two different sectoral approaches. On the one hand, human beings and their property have to be protected from natural hazards such as flooding and surges, and extensive agricultural use of the area has to be facilitated. Most of the times the development of an ecologically sustainable environment which is the aim of landscape planners and biologist is neglected.

There may be a common interest to preserve a nature resource, e.g. a National Park or Monument or relating to our topic a National Seashore or Lakeshore like Gulf Islands National Seashore or Indiana Dunes National Lakeshore in the USA.

The usual way of a planning process in floodplain area development is that a team of experts of just one of the two disciplines sets up a planning scenario which has to be assessed by the other discipline. For assessing the scenario, a set of additional requirements for the latter is developed and the file is returned to the first team, which has to integrate all the requirements into the planning scenario. This procedure is expensive and time-consuming and sometimes has to be reiterated a couple of times until a final compromise is found. For this reason, it is necessary to shorten the planning procedure and to develop an integrative planning process where all the available experts engaged in planning processes for floodplain areas are involved from the very beginning in order to save time, money and other resources.

Therefore, an integrative planning process has to meet at least the following requirements:

- identification of conflicts as early as possible in the planning process;
- documentation and presentation of the conflicts;
- possible alternatives for solving, compromising or mitigating the conflicts;
- consenting on a resolution or compromising of the extant conflicts;
- proposal for decision making;
- presentation of solutions.

The first and fourth requirement would demand the application of an adjusted scale of measurement for the evaluation of planning alternatives. Such a scale can only be derived through a thorough discussion between the disciplines. In trying to meet all these (and potential additional) requirements, really interdisciplinary co-operation has to be conducibly consummated. The rather 'logical' approach to enable conflict identification, documentation and presentation of resolutions would therefore comprise geographical information technology and cartographic visualisation.

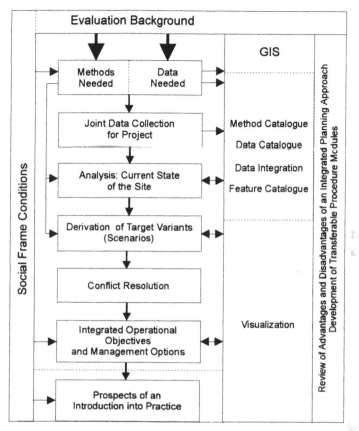

Figure 5.1 Reference model for an integrated planning methodology (after Lenk *et al.* 1997)

Concept for the integrated planning process

In the section above, the requirements for an integrated interdisciplinary approach to floodplain area development have been presented. A general schema of the reference model for this approach is shown in Fig. 5.1. First of all, it is necessary to establish a general evaluation background from which the data model and the required methods are being derived. The planning approach is based on the working hypothesis that there is a spectrum of objectives for floodplain area development which are located between the contrasting objectives 'protection of (or, alternatively, from; see below) use by man' and 'sustainable development of nature'. On the basis of the established methods not only the current state of the project area has to be assessed but also the objective variants being derived in certain planning scenarios. The GIS as an essential link between the pertinent disciplines plays an integrating role during the on-going communication and resolution of conflicts.

Concept of the GIS

The concept of the GIS is illustrated by Fig. 5.2. Due to the different views each discipline establishes its own primary model of reality, and describes the latter on the basis of geometric, semantic and temporal information including interrelations between these factors. This leads to the Digital Discipline-Specific Models (DDSM) of the subject areas, in this case models of the landscape planners and the civil engineers. The combination of DLM (see above) and DDSM leads to the Digital Object Model (DOM) which serves on the one hand as input for analysis and modelling calculations on the basis of methods implemented in the GIS, on the other hand it is used for the computer-assisted cartographic design process. The result of a cartographic design process leads to the Digital Cartographic Model (DCM) which is a virtual secondary model of the environment; it has to be transformed via DA-conversion to a suitable analogue secondary model of reality (Hake & Grünreich, 1994).

Derivation of an integrated data model

A central requirement for an interdisciplinary GIS-project is the development of a common data model, which includes all types of data needed by the project partners. The integrated feature catalogue represents this. Starting from the basic geodata feature catalogue given by ATKIS, the latter was investigated for correspondences and afterwards extended by defining the new types of objects to be integrated into the existing geo-data base.

Basic geodata set ATKIS. The Digital Landscape Model 25/1 (DLM25/1) as part of ATKIS is used as the basic geo-data set. It is readily available in its first implementation DLM25/1 from the surveying and mapping agencies of the federal states of Germany. ATKIS describes the landscape in an object-structured way from a topographical point of view. It consists of the 2-D Digital Situation Model 25/1 (DSM25/1) and the Digital Terrain Model 1:5.000 (DTM5) and is based on a hierarchical feature catalogue.

The latter classifies the environment into topographic objects of seven domains of feature classes; control points, settlements, transportation, vegetation, hydrography, relief and areas. These are again subdivided into several groups of feature classes, e.g. the feature class domain 'vegetation' is subdivided into the groups 'areas of vegetation' and 'individual trees and bushes'. The feature class groups then contain the actual feature classes such as 'individual tree' which are described further by various attributes, e.g. indicators for vegetation (see Fig. 5.3).

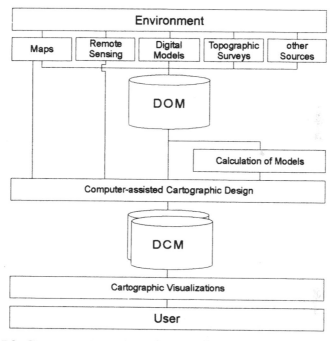

Figure 5.2 Concept of the GIS (after Hake & Grünreich, 1994)

ATKIS is designed to serve as a basic geodata set in order to provide other disciplines with basic information and a common reference frame for their discipline-specific data.

Required geodata. The information to be integrated into the planning model was chosen according to the requirements of data of the project partners and may be derived consistently from the evaluation background (see Fig.5.1). The contribution of the landscape planners to the planning model is based on the protected goods:

- species and biotopes;
- abiotic resources;

101

- experiencing nature and recreation.

as, for example, implied by the German Federal Law on Nature Conservation. These general data requirements have been supplemented according to the specific aspects of floodplain areas.

In Germany, the information needed for an operational water resource management is based on the German Federal Law on Water Resource Management and on regional laws, in this case the Law on Water Resource Management of the state of Lower Saxony. These laws demand that inshore waters are to be maintained in accordance to an operational state, but it is also necessary to take the concerns of nature management into account. This yields to the fact that any planting near or in the water has to be investigated with respect to its impact on the water budget. The protected good use of floodplains by man/protection of man-made goods has to be considered in particular. Therefore, it is necessary that simulation procedures are conducted to localise and assess any changes in water budget and transport. The respective parameters have to be included in the feature catalogue.

The information requirements were transferred to a catalogue of necessary data. It is divided into the components of:

- basic geodata;
- anthropological-biological data;
- physical factors;
- various user-specific individual factors.

The landscape planners evaluated a parameter-oriented description of the landscape where vegetation units, micromorphologic units and land-use were mapped. The civil engineers contributed data, which is needed for the calculation of hydraulic models of the area. After the analysis of the catalogue of necessary data and the ATKIS feature catalogue with regards to semantic correspondences, new feature class domains have been set up according to the general structure of ATKIS and were extended to fulfil the requirements of the planning approach. During the analysis of the feature catalogue of ATKIS and the user-specific feature catalogue it was, we think, convincingly confirmed that especially land use is highly correlated with topographic objects.

This integrated data catalogue still has to be supplemented with respect to results of analysis and assessments to be carried out as so far, no secondary data has been derived yet.

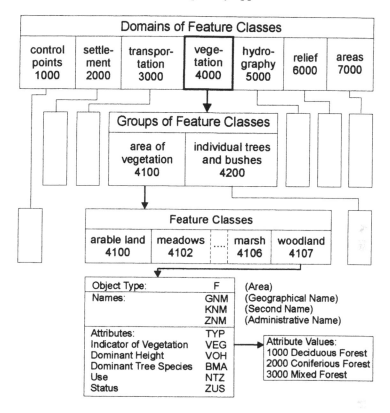

Figure 5.3 ATKIS feature catalogue (after Grünreich, 1992a)

Derivation of an integrated method base

Similarly as with the mentioned data requirements, an integrated method base must be developed as it is also important that all partners use in the first place the same set of methods for (e)valuation and assessment in order that results are unambiguous and reproducible.

In order to analyse and process the basic data as described in the data catalogue as an input for the evaluation methods, it is first of all necessary to establish a set of methods for data capture and prognosis. This includes among others the typification of land-use, vegetation units and micromorphologic units, the calculation of water transport, determination of ground water level, flooding duration and frequency of occurrence, as well as soil characteristics such as wind and water erosion and tendency to compression of the soil.

At the planning model in hand, the evaluation of protected goods is conducted following the procedure of ecological impact analysis using methods from practice and others, partly modified or self-developed methods. The classification of the shapes of parameters is based on ordinal scales for the individual process steps, and the delineation of results is based on matrix operations and verbal-descriptive methods. The methods may be divided into three sections, judged from the point of view of protection of 'species and biotops', 'abiotic resources' and 'land-use by man and possibility of use of resources'. One part of the project deals also with the general implementability of these methods in GIS.

Classification of the approach

Relating back to the sections above, where the different types of interdisciplinary approaches have been presented, it is now the aim to classify the approach given above.

According to the description of the project, we can highlight the following facts, which may be considered when assessing the level of interdisciplinarity of this project:

- three departments are working together on this joint approach to floodplain area development;
- there is a common data model satisfying the needs of the disciplines involved in the project;
- there is a common data base according to the data model;
- there is an integrated method base under development in order to make results of analysis unambiguous and reproducible;
- the people working on this project have different professional background.

The first fact states that three disciplines, which used to work separately, are working now together in a research area, i.e., it is an aggregated multidisciplinary field of research. This would relate the approach of being of level 3. The still inherent distinction between the three specific disciplines is even more demonstrated by the 5th fact stating that the people working on the project have different professional backgrounds, although we must say that at the time there is no real alternative to this circumstance. The latter argument denies in addition the classification of the approach of being of a genuinely interdisciplinary type (level 4). Additionally, this fact includes that so far no common theoretical integration has taken place. A step into this direction is given by the development of an integrated method base (see fact 4) giving reasons for the approach of being of level 5 of interdisciplinarity, nevertheless, a complete theoretical integration seems to be not finished yet.

A common data model and consequently a common data base is only the aggregation of data under a common logical and maybe a common semantic data model too. However, the data in the data base still can be interpreted in very different kind of ways with respect to the various interests and values applied to decision making. The latter may illustrated be the example of the stated case of the Bailly plant (Obermeyer & Pinto, 1994). Hence, a common data model relates to an approach of level 1 of interdisciplinarity.

As a result, the approach of the research project is considered to be between the levels 3 and 5 as so far, no definite classification to a specific level of interdisciplinarity can be made.

Concept of a GIS for the coastal zone

Similar to the concept for the GIS in floodplain area development, a concept of a GIS for the coastal zone is given in Grünreich (1995), see Fig. 5.4. For the purpose of a GIS for the coastal zone, the traditional separation between the topographic landscape models and hydrographic models must be given up and data integration from various kinds of sources must be conducted in order to describe the coastal zone in a semantically and geometrically consistent way. As for the common geodetic reference frame, it is advisable to choose international datums that provide the opportunity to extend regional approaches to international projects to emphasise the international relevance of coastal zone management (see above).

The approach of interdisciplinarity needs to be at least of the same level as of the project described above to satisfy the requirements of integrated coastal zone management (cf. Gubbay, 1990) and must be extended to cover the various areas of responsibility of the disciplines engaged in that process. Similar to the approach to floodplain area development, this concept still has to be supplemented with respect to value orientation, qualitative constraints and humanware etc. The simple approach of setting up an integrated GIS with a common data model (i.e. level 1 of interdisciplinarity; see above) is not sufficient for the purpose of a management system for a complex system such as the coastal zone.

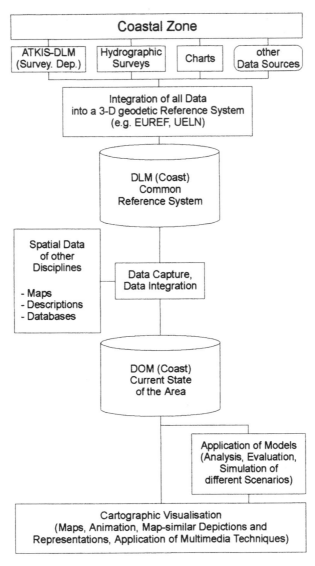

Figure 5.4 Concept of a GIS for the coastal zone (after Grünreich, 1995)

Similar research, as carried out for the case of a proposal for a UK coastal zone management plan (Gubbay, 1990), must be conducted to explore the responsibilities within the coastal zone in different countries bordering the area to be covered. The state of high-contaminated waters such as the North Sea is commonly known and must be improved on the basis of international co-operation. It is to be hoped that approaches to cross-boundary coastal zone management will be made possible through the members of the European Community.

Conclusion

Coastal zone development planning and management is a truly interdisciplinary enterprise of some complexity. It requires collaboration between experts and practioners from very different fields including public and military personnel, companies, scientists, particularly geographers, cartographers, geodesists, biologists, engineers and civil engineers specialised in water resource management, flood defence and maintenance, etc., planners, etc. Therefore, specific methodological and organisational requirements facilitating or even making possible interdisciplinary co-operation in the first place have to be met.

Geographical information technology may play a major role in conducting, structuring and controlling the respective coastal zone development and management operations within a wide-spread system. GIS is also interdisciplinary. Eventually, they as any further zone development and management have increasingly to take into account beyond the quantificative and purely geometric data bases also fundamental humanistic factors like value and interest orientations. Those may lead to value discrepancies and interest conflicts, which cannot be solved by the scientific GIS methods alone (only factual or 'cognitive' differences may be resolved). Therefore, alternative models of a rather societal character have to be developed in order to guarantee social acceptability and practical viability of proposed resolutions for developmental and managerial tasks of this rather encompassing systems type. In the future, to be conducive for special applications, even GIS methods themselves may be expanded to be able to comprise those sociogenic factors of embedding the respective planning methods into socio-technical, systems analytical, and ecological contexts.

'Humanware' in the mentioned sense (value commitments and problems, organisational and personnel as well as psychological and even motivational considerations) has to be supplemented to the existent GIS procedures or to be integrated in hypothetical expanded socio-systemic and ecological improved versions of GIS in a broader sense. Methodological requirements analysed by a special philosophy of socio-eco-technical systems approaches have to be met as well as organisational and leadership-oriented (e.g. motivation strengthening) preconditions conducive to socio-practically successful planning projects. All these conditions and criteria could only be sketched down here. They have to be worked out in detail in the future.

In this paper, we have tried to make explicitly the indispensability of elaborating and meeting (at least in general orientation and outlook) these conditions and criteria of interdisciplinary co-operation in a practice-oriented research as well as in project planning, management and actual operation. A specific example project of an interdisciplinary integrated floodplain area development based on GIS methods was used to highlight some of the modelling, data finding and data integration problems as well as the interdisciplinary 'link function' of GIS in this realm of research and planning (theory). In the end, a short outlook (after Grünreich, 1995 with extensions by the authors) is given to apply the described integrated modelling techniques based on GIS to coastal zone problems integrating traditional topographic and hydrographic approaches. This has still to be expanded to cover the human-oriented ('humanware'), societal and ecological factors mentioned.

References

Bunge, M., 1967. *Scientific Research, Vol. 1.* Springer, Berlin, Heidelberg, New York.

Campbell, H. & Masser, I., 1995. *GIS and Organisations.* Taylor & Francis, London.

ECDGXIII, 1997. GDF Home Page. URL: http://205.139.151.5/ehq/gdf/index.htm, visited 07.08.97.

Gottl-Ottlilienfeld, F., 1923. *Wirtschaft und Technik. Mohr-Siebeck,* Tübingen.

Grünreich, D., 1992a. *ATKIS - A Topographic Information System as a Basis for a GIS and Digital Cartography in West Germany.* Geol. Jb.

Grünreich, D., 1992b. Aufbau von Geoinformationssystemen im Umweltschutz mit Hilfe von ATKIS. In: Günther, O., Schulz, K.-P., & Seggelke, J. (Hrsg.): *Umweltanwendungen geographischer Informationssysteme,* 3-14. Wichmann Verlag, Karlsruhe.

Grünreich, D., 1995. *Stand der GIS-Technik und Anwendungen in der Hydrographie. In: Hydrographische Vermessungen - Heute - Schriftenreihe des Deutschen Vereins für Vermessungswesen,* 32-42, Konrad Wittwer Verlag, Stuttgart.

Gubbay, S., 1990. *A Future for the Coast? Proposals for a U.K. Coastal Zone Management Plan.* Marine Conservation Society, Ross-on-Wye.

Habermas, J., 1968. *Erkenntnis und Interesse.* Suhrkamp, Frankfurt a. M..

Hake, G. & Grünreich, D., 1994. *Kartographie.* Walter de Gruyter, Berlin, New York.

Hillmer, H., Peters, R. & Polke, M., 1976. *Studium, Beruf und Qualifikation der Ingenieure.* VDI-Verlag, Düsseldorf.

Huxhold, W. E. & Levinsohn, A. G., 1995. *Managing Geographic Information System*

Projects. Oxford University Press, Oxford.

IHB, 1993. *IHO Transfer Standard for Digital Hydrographic Data, Version 2.0.* International Hydrographic Bureau, Monaco.

Ingham, A. E. & Abbott, V. J., 1992. *Hydrography for the Surveyor and Engineer.* Blackwell Scientific Publications, Oxford.

Krüger, L., 1987. Einheit der Welt - Vielheit der Wissenschaft. In: Kocka, J. (Ed.): *Interdisziplinarität,* 106-125. Suhrkamp, Frankfurt a. M..

Lecher, K., Langer, H. & Grünreich, D., 1996. *Unterstützung des Planungsprozesses durch Geo-Informationssysteme bei der ökologisch orientierten Fließgewässer-planung.* Intermediate Report in Research Project at Hannover University, Germany. Hannover University, unpublished.

Lenk, H., 1970. *Philosopie im technologischen Zeitalter (2nd Ed.).* Kohlhammer, Stuttgart.

Lenk, H., 1977. *Team Dynamics.* Stipes, Champaign IL, U.S.A..

Lenk, H., 1983. *Eigenleistung.* Fromm, Zürich-Osnabrück.

Lenk, H., 1987. Werte als Interpretationskonstrukte. In: Lenk, H.: *Zwischen Sozialpsychologie undSozialphilosophie,* 227-237. Suhrkamp, Frankfurt a. M..

Lenk, H. & Maring, M., 1990. Responsibility for Land Use and the Problem of Social Traps. In: Fitch, D. B. S. & Pikalo, A. (Eds.): *Soziale und ökonomische Aspekte der Bodennutzung - Socio-Economic Aspects of Land Use Planning,* 31-50. Peter Lang, Frankfurt a. M., Bern, New York, Paris.

Lenk, H., 1993. *Interpretationskonstrukte. Zur Kritik der interpretatorischen Vernunft.* Suhrkamp, Frankfurt a. M..

Lenk, H., 1995a. *Schemaspiele: Über Schemainterpretationen und Interpretation-skonstrukte.* Suhrkamp, Frankfurt a. M..

Lenk, H., 1995b. *Interpretation und Realität.* Suhrkamp, Frankfurt a. M..

Lenk, H., 1997. *Interdisziplinarität und Interpretationskonstrukte.* (Paper submitted at Strasbourg University 1997, in press.)

Lenk, U., Grünreich, D. & Buziek, G., 1997. Interdisciplinary ecological planning of floodplain areas on the basis of ATKIS - Conception and first results. In: Hodgson, S., Rumor, M. & Harts, J.J. (Eds.): *Geographical Information 97, Volume 1, From research to application through co-operation.* Third Joint European Conference & Exhibition on Geographical Information. Vienna, Austria, 1997, 728-737, IOS Press, Amsterdam, Berlin.

MEGRIN, 1997. http://www.ign.fr/megrin/megrin.html, visited 10.08.97.

Munro, R.M. & Bradbury, R.H., 1996. *Using information technology for coastal zone management.* http://www.nric.gov.au/nric/publishing/papers/czerm1.html, visited 31.07.97.

MUV-B-W, 1996. Fachdatenintegration in ATKIS für das Umweltinformationssystem Baden-Württemberg (FDI-ATKIS-UIS). Ministry of Traffic and Environmental Affairs in Baden-Württemberg, Germany, Final Report in Research Project. Compiled By: Arslan, A., (FAW), Bannert, B., (IFK), Beuerle, R., (FAW),

Ebbinghaus, J., (FAW) & Grünreich, D., (IFK).

NOAA, 1997. *Coastal Zone Management Act of 1972*. http://www.nos.noaa.gov/ocrm/czm/CZM_ACT.html, visited 31.07.97.

Obermeyer, N. J. & Pinto, J. K., 1994. *Managing Geographic Information Systems*. Guilford Press, New York, London.

OII, 1997. *Geographical Information Standards (by Open Information Interchange Initiative)*. http://www2.echo.lu/oii/en/gis.html, visited 07.08.97.

Rimscha, S. V., 1996. Plain sailing into hi-tech harbour. *GIS Europe*, 5 (6): 30-32.

Saaty, Th. L., 1980. *The Analytic Hierarchy Process: Planning, Priority Setting, Resource Allocation*. McGraw-Hill, New York.

Scheuermann, W., 1997. *What is ECDIS and what can it do?* http://lobby.sevencs.com/7cs/aboutECDIS/ecdis_en.html, visited 07.08.97.

Stough, R. R. & Whittington, D., 1985. Multijurisdictional waterfront land use modeling. *Coastal Zone Management Journal*, 13, 151-175.

Torge, W., 1991. *Geodesy*. Walter de Gruyter, Berlin.

Walter, V., 1997. Zuordnung von raumbezogenen Daten - am Beispiel der Datenmodelle ATKIS und GDF. *Geo-Informations-Systeme*, 10(2): 3-4.

Wang, M. & Stough, R. R., 1986. Cognitive analysis of land-use decision-making. Modeling and Simulation. *Proceedings of the 17th Regional Science Association Conference in Pittsburgh, Part 1: Geography-Regional Sciences, Economics*, 17, 107-112.

Woodsford, P., 1995. *The Significance of Object-Orientation for GIS*. http://www.laserscan.com/papers/ooforgis.htm, visited 05.08.97.

Acknowledgements

The authors are grateful to the colleagues from the departments of Hannover University engaged in the project described above, Dr.-Ing. G. Buziek of the Institute of Cartography; Prof. Dr. sc. techn. K. Lecher and Mr Dipl.-Ing. C. Lange of the Institute of Water Resources, Hydrology and Agricultural Hydraulic Engineering and Prof. Dr. rer. nat. H. Langer and Ms Dipl.-Ing. B. Knickrehm, Institute of Landscape Planning and Nature Conservation.

6 The total economic value of the natural resources of the coast

S. GOODMAN, W. SEABROOKE AND S. JAFFRY

Introduction

Increasingly, development agencies are attempting to consider the total economic value of natural resources in their economic assessments of policy options. Total economic value consists of values derived partially from use and partially from the discrete existence of natural resources taking a naturally determined time horizon. Over the centuries, most societies have created institutions, which facilitate and control the use of natural resources for human benefit. For example, market mechanisms to allocate scarce resources between competing users and systems of private property rights to identify use rights.

Valuers and economists are relatively confident of their ability to measure the price which users are prepared to pay to exercise their right to use natural resources. The value of benefits derived from use is invariably linked to a finite time horizon, seldom greater than 20-25 years. There are some areas of difficulty, for example, where use benefits are modest in relation to the transaction costs associated with them, but by and large, economists have found ways of coping with such problems. Valuations based on use pay greater regard to financial erosion than to the physical erosion of resources unless required to do so by some powerful reason (such as statute).

In contrast to this utilitarian view of natural resources, the cultural tradition of depicting the natural environment has, almost universally, emphasised intrinsic qualities rather than use characteristics: the freshness of the air, the clarity of water, the richness and variety of habitat. The value frameworks used to depict these characteristics are sometimes aesthetic - depicted in artistic and literary images sometimes scientific, but never economic. Hitherto, the value of these intrinsic characteristics has tended to be acknowledged in economic terms only when their existence has become threatened or eroded.

Increasingly, however, the general public is are being challenged to quantify its preference for some of these things - just how much are we prepared to pay for pure, uncontaminated water; how much are we prepared to pay to minimise health hazards? This, of course, starts to raise the question of the economic value of the intrinsic qualities of the terrestrial and marine environment. These features of our shared environment have tended to be treated as being discrete from day to day economic existence rather than integrated into it. The sophisticated social institutions of day to day existence have left this area untouched other than posting a rather feeble 'keep off' sign. Rent seeking behaviour encourages the most economically voracious or careless individuals and corporations to trespass into these areas, in the knowledge that the sanctions against them are relatively weak.

There is no doubt that existence values do exist, we simply lack robust and rigorous economic mechanisms to recognise them and incorporate them into value models dominated by use values. As development agencies are forced to consider the total economic value of natural resources in their economic assessments of policy options, attention is turning to methods for including existence values into the economic appraisal of projects which have any impact on our natural environment. The research on which this paper is based addressed the natural coastal environment of the UK.

It is worth highlighting a number of economic assumptions to preface this paper:

- the value of natural resources derives from the benefits people receive from them over a naturally determined time horizon;
- scarcity forces choices to be made between competing resource uses;
- economics provides a framework for evaluating the costs and benefits of each choice in the simple, consistent and almost universal dimension of 'money';
- the total economic value of natural resources is maximised when they are used in ways that yield the greatest sum of net individual benefits.

Of the methods developed to estimate total economic value, currently, the most widely used technique for evaluating the value of natural resources is the Contingent Valuation Method (CVM). (See Mitchell & Carson (1989) for a complete discussion of the GVM). The CVM provides a framework for estimating non-use values in that it captures user benefits (use and non-use) and non-user benefits. However, CV studies are too expensive and time-consuming to justify their use for every project appraisal. The major studies have tended to be geographically specific and the sites sufficiently unique that it is difficult to derive from them generic information that would form a sound basis for constructing values for less unique sites. The utility of the CVM is impaired by lack of a cost- effective means of applying the technique to a wide range of natural resources. This severely restricts the viability of CVM for routine policy evaluations (Mitchell & Carson, 1989; Harrison & Lesley, 1996).

Policy-makers are more interested in interpreting benefits estimated from one study to other, comparable sets of conditions. This is known as the 'Benefits Transfer' approach and is based on secondary analysis, taking the results of primary data and applying them to comparable examples. Integrating an assessment of total economic value into routine policy appraisals requires a benefit- transfer approach, which is properly designed and applied yet which is cost-effective. However, securing a consistent basis for estimating the economic value for all benefits is not easy.

In order to develop a rigorous basis for estimating the total economic value of a wide range of coastal resources, this study explored the use of a conservationist valuation framework to assess preferences for stereotypical features of the British coast. This framework identifies and provides a relatively objective basis for evaluating the physical characteristics that give natural resources a conservation value and identifies criteria that are widely accepted as being significant in determining conservation value. In contrast to other benefit flows such as aesthetics or landscape values, conservation values are more amenable to objective assessment.

As an initial step in the development of a modelling framework linking conservation and economic values for natural resources, a CV survey was designed to evaluate the extent to which the economic value of conservation quality corresponded to conservationists' (non-monetary) assessment of the conservation quality of coastal resources. The primary objectives of the study were to evaluate:

- the relationship between conservation and non-use values for

coastal resources;
- the relationship between scientific and public assessments of conservation quality;
- the ability of the CVM to estimate conservation values in economic terms. (For the full report see Goodman, S. *et al.* 1996)

Hypothesis

A direct relationship between non-use and conservation values would support the hypothesis that conservation quality levels could be used as a surrogate in the attribution of non-use values to coastal resources. Conservation values can be estimated for a broad range of site characteristics, including those, which are not considered to be of particularly outstanding quality i.e. ubiquitous rather than unique. If the relationship proved to be sufficiently robust, then non-use values could be benchmarked to conservation values. This study sought to establish the existence of such a relationship.

The CV survey was designed to determine whether or not individuals' willingness to pay (WTP) to protect the conservation quality of coastal areas depended upon the current level of conservation quality. Thus, the survey tested the null hypothesis that

$$WTP_{hq} = WTP_{lq}$$

where the suffix $_{hq}$ denotes a conservation programme in coastal areas of high conservation quality, and the suffix $_{lq}$ denotes a conservation programme in coastal areas of low conservation quality.

Methodology

Questionnaire design

Information about the characteristics of the British coast was given to respondents on a showcard. This information had to reflect the views of conservationists yet be readily understood by the average respondent. It was formulated in conjunction with environmental scientists from the Ministry of Agriculture, Fisheries and Food (MAFF), English Nature, and the

National Rivers Authority.

This group of scientists agreed that the English coast could be characterised by five major landforms, namely:

- hard rock cliffs and headlands;
- soft earthy cliffs;
- sand and shingle spits and bars;
- sand dune systems;
- estuaries.

Information regarding the conservation characteristics of coastal habitats was based on the assessment of this expert panel that the following criteria are most important in creating conservation value across the range of habitats in accordance with the Ratcliffe criteria (Ratcliffe, 1977):

- area of habitat;
- species diversity;
- species populations;
- freedom from disturbance.

Consequently, the characteristics that give these landforms conservation value were determined to be:

- diversity and rarity of habitats and species;
- ecological specialisation;
- freedom from disturbance;
- size/lateral extent; and
- dynamism of geomorphological processes.

The showcard described a hypothetical management programme designed to maintain current levels of conservation quality along the entire coast. It stated that, without the programme, coastal areas would experience what scientists considered to be a 'critical' level of loss, which was defined as a 75% loss of plant and animal species and populations and a loss of coastal habitats which could not be reversed over a 30 year period. The loss of conservation value was evaluated over a human generation largely because of this period's relevance for policy-making purposes. *This Common Inheritance* (HMSO, 1990) has suggested that the time-span of a human generation is appropriate for evaluating environmental sustainability. English Nature (1992) has, in turn, suggested that the theory of environmental sustainability be operationalised by

> identifying elements of the natural environment whose loss would be serious, or which would be irreplaceable, or which would be too difficult or expensive to replace in human time scales.

As far as possible, the questionnaire was designed to conform to guidelines concerning 'best practice' CV studies, which was presented to the U.S. National Oceanic and Atmospheric Administration (NOAA) (Arrow, *et al.*, 1993) by a panel of renowned economists. This panel expressed particular concern over embedding effects, or part-whole biases when evaluating complex environmental goods. These occur when respondents are insensitive to the extent of environmental protection being considered, and perceive a good other than that which they are asked to value. Such biases distort the welfare benefits estimated through the GVM.

The study on which this paper is based incorporated measures to minimise embedding effects, drawing on good practice established in other studies (Kemp & Maxwell, 1992; Green *et al.*, 1991; Bateman *et al.*, 1992; Garrod & Willis, 1996.) Respondents were asked to disaggregate any budget they might wish to allocate to a coastal conservation programme from:

- their overall budget for environmental protection;
- their budget for a conservation programme in a limited group of coastal areas.

(They were reminded that other coastal areas existed where a conservation programme could be started and that money spent on a conservation programme in a specific group of areas would not be available for spending elsewhere.)

The study also adopted the split-sample, single-good format to minimise embedding effects (see Mitchell & Carson, 1989; Imber, *et al.*, 1991; Carson *et al.*, 1995) and adopted Willingness-to-Pay (WTP) through the medium of additional taxes as the format for the expression of respondents' preferences.

(The NOAA panel recommend a discrete/dichotomous choice (DC) or referendum format for eliciting WTP. Support for the DC format is based largely on the assumptions that it is 'incentive compatible' (Hoehn & Randall, 1987) and more closely reflects market decisions than open-ended valuation questions. However, a growing body of research challenges these assumptions and, therefore, the preference for the DC format. Studies in experimental economics have found that DC expressions of value consistently overestimate demand-revealed values whereas open-ended questions provide a better approximation to revealed values evidenced by auction prices (Smith & Walker, 1993; Balistreri *et al.*, 1995). Furthermore, research from cognitive psychology indicates that when asked to evaluate a good which is unfamiliar or complex, respondents are inclined to interpret any suggested price as an indication of its true value (Kahneman, *et al.*, 1982; Harris, *et al.*, 1989). These findings are supported by CV studies which have noted the tendency for respondents to DC questions to say they would pay the bid with which they are presented to signal their support for the environmental asset under evaluation (Whitehead, *et al.*, 1995; Bateman, *et al.*, 1995). For these reasons, and supported by the findings of CV studies which have shown that open-ended WTP questions tend to yield lower benefit estimates than the DC format (see for example, Bishop *et al.*, 1994; Brown, *et al.*, 1996), the former was adopted as a more conservative design strategy, producing a lower bound estimate of respondents' true VVTP.)

The valuation scenario began by establishing the 'payment-principle', asking respondents if, in principle, they would be willing to pay additional taxes for a conservation programme for the coast of Britain. Those who responded positively were then asked to specify the maximum amount of additional tax, which they would be willing to pay for this programme.

(To increase the reliability of estimated WTP, respondents were shown a tax booklet listing the amount of taxes a married couple at the mid-point of various income ranges currently paid for a variety of public services. This technique has been used by other CV researchers to reduce the random error associated with open-ended WTP questions (e.g. Mitchell & Carson, 1989; Willis, *et al.*, 1993). However, it must be noted that any increase in reliability is achieved at the cost of in creased bias. Several respondents' comments indicated that they anchored their WTP for a coastal conservation programme to the amount of tax paid for other public services. Whilst earlier studies (Bateman, *et al.*, 1992; Willis, *et al.*, 1993) and our own pre-testing found that increased taxes are generally perceived to be a neutral payment vehicle, the central government budget was under debate at the time this survey was conducted. Unfortunately, taxation policies appear not to have been neutral topics in the final months of 1995.)

Hanemann (1991) and Tversky *et al.* (1990) have shown that WTP to avoid a loss of environmental quality differs significantly from WTP to gain an improvement in environmental quality. Thus, care was taken to assure that respondents understood that the potential benefits to be gained from a coastal conservation programme would arise from avoiding a loss of conservation quality, not from increasing conservation quality levels.

Respondents who stated a positive tax amount for the coastal conservation programme were asked how much of this additional tax they would like to see spent in one of two quality categories (high or low). Group 1 represented areas with a relatively high conservation quality, and Group 2 characterised areas with a low conservation quality. Respondents were given information about either Group 1 or Group 2. (Interviewers rotated showcards between each interview). In order to focus on non-use values it was desirable to direct respondents' thoughts away from their experiences with specific areas of coast and toward coastal conservation characteristics. Respondents were given stereotypical descriptions expressed in terms of the five coastal landforms and their key conservation characteristics depicting the relevant conservation quality for each group. The showcard stated that each group represented about 10% of the coastline of Britain and described the conservation value of the remaining 90% of the coast in relation to these areas. Pre-testing had indicated that 10% of the coast was large enough to be significant to respondents yet small enough to be plausible in a descriptive sense.

Interviews with respondents who were unwilling to express a WTP for coastal conservation provided no economic information which could be used to estimate the social benefits of coastal conservation quality. Nevertheless, in view of the other objectives for this study, namely, to investigate public perceptions of coastal conservation quality and to compare public and scientific assessments of conservation quality levels, it was important to question respondents about their perceptions of conservation quality. Those who were unwilling to pay additional taxes for a coastal conservation programme were told that, even without additional taxes, a conservation programme could be started in selected coastal areas. They were given a showcard, which described the two groups of coastal areas. The showcard also reminded respondents of other coastal areas where a conservation programme could be started, and stated that the remaining 80% of the coastline consisted of areas with levels of conservation quality lying between the high and low levels described.

The questionnaire included follow-up questions to investigate the reasoning behind respondents' WTP statements as well as questions to determine how respondents reacted to the valuation scenario. Information was also collected on coastal usage patterns and the socio-economic profile of each respondent household so that a theoretical validation model of WP responses could be developed. Following recommendations of the NOAA panel, the questionnaire allowed for respondents' ambivalence, uncertainty, or need for additional information by incorporating a no-answer option.

Sampling and interviewing strategies

Central England was chosen to increase the likelihood of interviewing coastal non-users. The survey was conducted through personal interviews from November 1995 through January 1996. A sampling frame of 2000 addresses was created by extracting a stratified sample of in the vicinity of the city of Derby. Enumeration districts were stratified according to a national index of socio-demographic characteristics and selected proportional to population. Respondent households were randomly sampled from each selected enumeration districts and mailed a letter requesting their participation in a survey of 'environmental and other issues' being conducted by the University of Portsmouth. This resulted in 806 interviews being conducted, structured to ensure that a representative cross-section of the adult British population was sampled. As a result, nearly 95% (766) of the 806 interviews produced usable questionnaires.

Results

The relationship between conservation and non-use values

One of the difficulties with estimating discrete non-use values is that there is no theoretical basis for disaggregating an assessment of total value into use and non-use components. For this reason, non-use values are ideally estimated by questioning individuals who do not use a resource about their value for it. However, despite the selection of a sampling locus intended to increase the chances of interviewing coastal non-users, 98% of our sample had visited the coast, with over 75% having done so within the past year. Nevertheless, while it is not realistically possible to select, from within the UK, a sample of non-users of the British coast that is representative of the population as a whole, the approach of evaluating preferences for conservation characteristics does appear to be effective in capturing non-use benefits for the natural coastal environment. The value estimates generated by this study can therefore best be interpreted as estimates of respondents' total value (which includes any relevant non-use values) of the conservation quality of the British coast.

Analysis of the reasons stated by respondents for protecting natural resources indicated that a substantial portion of their value of environmental resources is related to non-use motivations. Almost 90% of respondents stated that non-use motivations (i.e. existence, bequest, and option values) were extremely or very important reasons for protecting natural resources. Responses to open-ended questions also indicated that respondents value the conservation quality of the coast of Britain for reasons that are, in large part, unrelated to its use. In addition, many of the reasons respondents gave for wanting part of their coastal conservation programme budget to be spent in Coastal Groups I and II indicated the influence of non-use values. Statements relating to the need to conserve coastal habitats and wildlife, maintain current levels of conservation quality and prevent further losses, and protect all coastal areas equally strongly, suggest that part of respondents' value of coastal areas is related to their intrinsic (i.e. existence) value.

From a total response of 766 usable interviews, 76% of respondents said that, in principle, they would be willing to pay additional taxes for a coastal conservation programme. Of these, approximately 5% subsequently stated a maximum WP of zero, and a similar percentage remained unsure, throughout the survey, about how much additional tax they would be willing to pay for the programme and failed to state a WTP amount. This resulted in a total of 69% (528) of respondents stating a positive amount of additional tax, which they would be willing to pay for the programme.

The validity of WTP bids was evaluated by investigating respondents reasons for agreeing to pay for a coastal conservation programme in one of the two groups of coastal areas. Nearly 83% of all WTP bids were considered valid estimates of respondents' true WTP. Almost 17% of the bids indicated that they had been made to show their support for environmental issues in general and were considered invalid bids. Eight bids exceeding 150 were considered to be extremely high. Further analysis indicated that three were valid. Of the remaining five invalid bids, three suggested strategic bidding. Overall, protest behaviour in the form of strategic bids and free-riding does not appear to have been a problem in this study.

When asked to allocate their total budget for a coastal conservation programme to the coastal group which they evaluated, 34 respondents either did not want any of their budget to be spent in that group or did not know how much of this budget they wanted to spend in this group. The reasons behind these zero and 'not sure' responses indicated that these respondents either had other spending priorities, did not consider themselves qualified to make such a decision, or did not have enough information to state their choice. These bids were excluded from calculation of mean (positive) WTP. The result of this approach would be to increase mean WTP, and overestimate any aggregated benefit estimates. This effect is counter-balanced by expressing aggregated positive mean WTP as a percentage of the entire sample.

For several reasons some respondents were unable or unwilling to state their willingness to pay additional taxes for a coastal conservation programme. These included:

- some might not value the conservation quality of the coast and, therefore, would not benefit from a conservation programme;
- some might not be able to afford paying for the programme;
- some might value the conservation quality of the coast but refuse, for moral reasons, to make economic trade-offs in favour of environmental quality;
- some believe that the value of environmental resources is intrinsic, so no increase in income would compensate them for a reduction in environmental quality (Stevens, *et al.*, 1991; Hanley, *et al.*, 1995);
- some might have strong feelings and values for natural resources but are unable to express their values in monetary terms (Gregory *et al.*, 1993);
- some could have rejected, or had problems understanding, some aspect of the valuation scenario.

The reasons were investigated to determine their consistency with economic behaviour:

- 6.5% (50) of respondents stated that they could not afford to pay additional taxes or that they would not benefit from the

121

programme, indicating that the value of the programme to them was zero;

- 21% (161) of respondents objected to some aspect of the valuation scenario, notably, 121 objected to higher taxes;
- Approximately 3.5% (27) of respondents said that they needed more information to value the programme;
- 1% (12) objected to having been asked the question;
- Less than 1% (9) responses indicated free-riding behaviour.

Respondents who answered 'don't know' or 'other' were asked to specify their uncertainty and other reasons. Analysis of these responses yielded five additional valid zero bids and 14 additional tax protest bids.

Respondents reported a mean (positive) WTP of 48.36 per annum in additional household taxes for a coastal conservation programme for the English and Welsh coast. The median value was 25 per annum. Of the respondents who evaluated Coastal Group I (high conservation quality), the average amount nominated to be spent on high quality sites from their total budget for a coastal conservation programme was 24.75. The corresponding amount nominated by respondents evaluating Coastal Group 11(10w conservation quality) was 17.87. The median value for a conservation program in each coastal group was identical at 10.

Statistical tests were conducted to uncover any embedding effects, which might distort the estimated benefits derived from coastal conservation quality. The mean of respondents' estimated annual household expenditures on environmental protection was 21.84. Univariate statistical testing (paired t-tests) showed that mean and median annual household expenditures on environmental protection were significantly different from mean and median household WTP for a coastal conservation programme at the 0.05 level. This indicates that respondents' value of the conservation quality of the British coast was not embedded in a broader concept of environmental protection in general.

Similarly, a t-test comparison showed that the difference between mean and median WTP for a conservation programme for the entire coast and a corresponding programme applied to 10% of all coastal areas was significant at the 0.05 level. This indicates that respondents appropriately distinguished the extent of environmental benefits provided by varying levels of the conservation programme's inclusiveness, and suggests that perfect embedding did not occur at this level of analysis.

Univariate analysis revealed that the difference between means for Coastal Group I (high quality) and Coastal Group 11(10w quality) was significant at the 0.10 level, but not at the 0.05 level. Multivariate analysis of the difference in mean WTP amounts was also conducted. This is a preferable indicator of WTP differences as it considers the effects of other variables affecting WTP, as well as an area's level of conservation quality. A dummy variable representing area category' was included in the valuation function modelling WTP for a conservation programme in a limited group of coastal areas. This variable was not significant at the 0.10 level. (The dummy variable representing 'area category' measured any shifts in the demand curve. Dummy variables measuring changes in the slopes of the demand curves for the two groups of coastal areas were also found to be not significant at the 0.05 level.) The lack of a significant difference between mean WP for a conservation programme in Coastal Groups I and II indicates that, overall, respondents did not express an economic preference for higher, rather than lower, levels of conservation quality.

The weak significant difference between mean WTP for the two groups of coastal areas suggests that either respondents did not hold such a preference, or they were unable to express their preferences in monetary terms. The latter interpretation is supported by the finding that over 54% of respondents who were not willing to pay additional taxes for a conservation programme for the entire coast selected Coastal Group A (high quality) as a priority for this programme, whilst only 24.5% selected Coastal Group B (low quality). While inconclusive, this suggests that, generally, respondents preferred higher levels of conservation quality but that it may have been difficult for some to express their preferences in monetary terms.

Valuation functions were modelled to assess the (theoretical) validity of the WTP estimates. These valuation functions were specified according to four variable sets, which economic and behavioural theories and common sense suggest should be positively related to an individual's willingness to pay for a coastal conservation programme. These variable sets consisted of:

- measures of respondents' environmental attitudes and behaviour;
- socio-economic status;
- use of the coast;
- familiarity with coastal conservation characteristics (understanding of CV scenario).

In the absence of an *a priori* assumption that the distribution throughout the sample of respondents who expressed a willingness to pay additional taxes and those who did not would be other than random, a sample selection univariate probit model was specified to predict the probability of a positive response to the 'payment-principle' question. (The problem of sample selection bias will occur if the probability of obtaining a valid WP response among respondents having a particular set of characteristics is related to their value for a good. The LIMDEP Version 7 sample selection univariate probit model was used to address sample selection bias. The alternative of using weighting and imputation procedures separately to estimate a probit model for the payment principle question and using two ordinary least square models for the two WTP questions does not address sample selection bias, and would produce biased results.) This model correctly predicted over 90% of responses to this question. In addition, it generated a variable (A), which was included as an explanatory variable in the remaining two valuation functions to test for sample selection bias.

Household income was the most important positive determinant of WTP for a coastal conservation programme. Membership in an environmental organisation had a positive effect on WTP for a conservation programme for the entire coast, and indicators of respondents' environmentally- supportive attitudes were important in explaining WTP for this programme along the entire coast, as well as at a limited group of coastal areas. Respondents who, in the interviewer's assessment, were relatively uncooperative at the end of the interview and who appeared to have given only superficial consideration to the WTP questions were found to be willing to pay significantly less than average for the coastal conservation programme.

Respondents' use of the coast appeared to exert conflicting effects of WTP. A recent visit to the coast (within the last six months) by any household member had a significant positive effect on WTP in both models, whilst frequent coastal visits (at least once every two months) by a household member had a negative impact on WTP for a conservation programme in a specific group of coastal areas. One possible explanation of this finding is that frequent coastal visitors may have had strong sentiments about protecting coastal areas with which they were most familiar, rather than the group of sites presented to them during the interview.

Explanatory variables from each 'core' variable set explained 19% and 11% of the variation in WTP for a conservation programme for the entire coast and this programme at a limited group of coastal areas, respectively. These results are similar to those of other CV studies conducted in the UK and US (see Mitchell & Carson, 1989; Bateman, *et al.* 1992; Willis, *et al.*, 1993). More importantly, from the standpoint of assessing the theoretical validity of our WTP estimates, all three valuation functions were consistent with theoretical and *a priori* expectations, as well as with one another.

Further evidence of the theoretical validity of the WTP estimates is provided by observing changes in mean WTP as variables, which are expected to influence WTP change. These variables include evidence of environmentally-supportive behaviour, coastal usage patterns, age, and household income. As expected, respondents from households where someone belonged to an environmental organisation expressed a higher mean WTP for a coastal conservation programme than respondents from households that did not belong to such an organisation. Similarly, recent coastal users were willing to pay significantly more than average for a coastal conservation programme. WTP for a coastal conservation programme also increased with income, although not continuously over all income ranges. As expected, WTP exhibited a quadratic relationship with the respondent's age, increasing up to middle age, and then decreasing.

Respondents' comments suggest that a substantial portion of their value of the British coast is related to its conservation quality. Additionally, existence, bequest, and option values appear to be significant components of their total value of the coastal environment. Their assessments of conservation quality levels were generally consistent with those of environmental scientists. The majority of respondents indicated that their perceptions corresponded to the 'scientists' distinction between high and low conservation quality in the groups of sites, which they were respectively asked to evaluate.

Respondents' reasons for wanting part of their conservation programme budget to be spent in a particular group of areas reveal that, to a large extent, individuals used different criteria to evaluate the threatened loss of conservation quality in each group. Reasons given for starting a conservation programme in the 'high quality' group tended to relate to protecting the high levels of conservation characteristics which respondents perceived within these sites. However, respondents who evaluated the 'low quality' group often perceived these areas to be more vulnerable and in greater need of a conservation programme for protection.

The greatest number of comments concerning the need for a conservation programme in the 'low quality' group of sites mentioned an aversion to coastal erosion and a corresponding need for sea defences. These comments highlight a difference between public and scientific assessments of coastal conservation value. From the perspective of nature conservation, areas where the natural processes of erosion are allowed to occur have higher value. However, some members of the public hold strong beliefs about coastal erosion which conflict with scientific thinking, which is currently influencing public policy. This finding is consistent with Green & Tunstalls (1991) finding that, while there is often a reasonable degree of congruence between scientific and public perceptions of ecological value, what the public values as a desirable environment does not necessarily coincide with the ecologically preferable management strategy for that habitat.

The existence of strong preconceptions about coastal processes, which may have overpowered new information presented to respondents during the interview, has important implications for interpreting the benefit estimates produced through the GVM. Work by Fischhoff & Furby (1988) has found that people often misinterpret or ignore new information that involves concepts, which are unfamiliar or difficult to grasp. Research by Gregory & MacGregor (1990) shows that problems arise for the CVM when respondents' evaluation of an environmental change requires a fundamental shift in their vision of the world for the change to be fully processed and understood. Our finding that some individuals selectively processed the information, which was presented to them and failed to consider the conservation benefits of erosion, is consistent with their findings

Aggregated benefits of coastal conservation quality

To be useful for purposes of policy evaluation, the estimated benefits, which are produced through the CVM must be aggregated across a relevant population. Within the current application, the total social value of the conservation quality of the British coast would be estimated by multiplying mean (positive) WTP of 48.36 by the appropriate percentage of British households. However, this approach would almost certainly overestimate the benefits derived by the British public from coastal conservation quality.

A growing body of evidence suggests that hypothetical values overstate 'true' economic values. This highlights the need to calibrate expressed preferences values in terms, which more accurately reflect revealed preferences. Several studies have found a wide disparity between the number of people who say that they would be willing to pay for goods and the number of people who ultimately do pay (see Coursey, *et al.*, 1987; Neill, *et al.*, 1994). The NOAA panel recognised that hypothetical values tend to overstate 'true' WTP for both private and public goods, and noted the need to calibrate hypothetical values before aggregating them. In their review of eight years of experimental studies, Balistreri *et al.* (1995) found that, on average, hypothetical bids exceed market values by a factor of 1.65. (In the absence of better information, NOAA recommends that WTP values derived through the CVM be divided in half in order to correct for the upward bias in hypothetical value statements [NOAA Workshop, 1994]). Researchers in the U.S. have found a difference as large as an order of magnitude between the number of people who state a willingness to pay higher utility bills (to develop renewable sources of electricity) and the number of people who actually support such pricing policies when they are implemented (Byrnes, 1996). The extent to which hypothetical WTP responses need to be calibrated to reflect more accurately 'true' WTP is currently unresolved amongst researchers and public agencies basing policy decisions on the results of CV studies.

Acknowledgements

The research on which this paper is based was commissioned by the Flood and Coastal Defence division of the Ministry of Agriculture Fisheries and Food whose support we gratefully acknowledge.

References

Arrow, K., Solow, R., Portney, P.R., Learner, E.E., Radner, R. & Schuman, E.H., 1993. Report of the NOAA Panel on Contingent Valuation. Report to the General Counsel of the US National Oceanic and Atmospheric Administration. US Department of Commerce, NOAA. Natural resource damage assessments under the Oil Pollution Act of 1990, 58 *Federal Register*, 4602-4614.

Balistreri, E., McClelland, G., Poe, C., & Schulze, W., 1995. *Can Hypothetical Questions Reveal True Values? A Laboratory Comparison of Dichotomous Choice and Open-Ended Contingent Values with Auction Values*, unpublished paper, Department of Economics, University of Colorado, Boulder, CO.

Bateman, I., Langford, I, Turner, R.K., Willis, K. & Garrod, G., 1995. Elicitation and truncation effects in contingent valuation studies, *Ecological Economics*,

12, 161-179.

Bishop, R.C., Welsh, M.P., Heberlein, T.A., 1994. *Some Experimental Evidence on the Validity of Contingent Valuation*, Department of Agricultural Economics, University of Wisconsin, Madison, June.

Brown, T.C., Champ, P.A., Bishop, R.C., & McCollum, D.W., 1996. Which response format reveals the truth about donations to a public good? *Land Economics*, **72**(2), 152-166.

Carson, R., Flores, N. & Hanemann, W.M. 1995. *On the Creation and Destruction of Public Goods: The Matter of Sequencing. Discussion Paper 95-2 1*. Department of Economics, University of California, San Diego.

Coursey, D.L., Schulze, W.D., & Hovis, J.J. 1987. The disparity between willingness to pay measures of value, *Quarterly Journal of Economics*, 680-690.

Edwards-Jones, G., Edwards-Jones, E.S., & Mitchell, K., 1995. A comparison of contingent valuation methodology and ecological assessment as techniques for incorporating ecological goods into land-use decisions, *Journal of Environmental Planning and Management*, **38** (2), 215-230.

English Nature, 1992. *Strategic Planning and Sustainable Development: An Informal Consultation Paper*. English Nature, Peterborough.

Fischhoff, Baruch, & Furby, 1988. Measuring values: a conceptual framework for Interpreting transactions with special reference to contingent valuation of Visibility, *Journal of Risk and Uncertainty*, 147-184.

Goodman, S., Seabrooke, W, Daniel, H., Jaffry, S., & James, H., 1996. *Determination of Non-use Values in Respect of Environmental and Other Benefits in the Coastal Zone. Results of a Contingent Valuation Study of Non-use Values for Coastal Resources*. Research Report to the Ministry of Agriculture, Fisheries and Food, Flood and Coastal Defence Division.

Gregory, R., Lichtenstein, S., Brown, T.C., Peterson, G.L., & Slovic, P., 1995. How precise are monetary representations of environmental improvements? *Land Economics*, **7**(4), 462-473.

Gregory, R., Lichtenstein, S., & Slovic, P., 1993. Valuing environmental resources: a constructive approach, *Journal of Risk and Uncertainty*, **7**, 177-197.

Gregory, R., & MacGregor, D., 1990. Valuing changes in environmental assets. In: Johnson, R.L., & Johnson, G.V., (eds.), *Economic Valuation of Natural Resources: Issues, Theory and Applications*. Westview Press, Boulder, Colorado.

The Environment White Paper, 1988. *This Common Inheritance*. Cmnd.1200. HMSO, London.

Hanemann, W.M., 1991. Willingness to pay and willingness to accept: how much can they differ? *American Economic Review*, June, 635-647.

Hanley, N., Spash, C., Walker, L., 1995. Problems in valuing the benefits of biodiversity protection, *Environmental and Resource Economics*, **5**, 249-272.

Harris, C.C., Driver, B/Q., & McLaughlin, W.J., 1989. Improving the contingent valuation method: a psychological perspective, *Journal of Environmental Economics and Management*, **17**, 213-229.

Harrison, G.W., Lesley, J.C., 1996. Must contingent valuation surveys cost so much? *Journal of Environmental Economics and Management*, **31** 226-247.

Hoehn, J.R., & Randall, A., 1987. A satisfactory benefit cost indicator from contingent valuation, *Journal of Environmental Economics and Management*, **17**, 226-247.

Imber, O., Stevenson, G. & Wilks, L., 1991. *A Contingent Valuation Survey of the Kakadu Conservation Zone.* Research Paper No. 3 (February). Resource Assessment Commission: Canberra, Australia.

Kahneman, D., Slovic, P., & Tversky, A., (eds.) 1982. *Judgment Under Uncertainty: Heuristics and Biases.* New York: Cambridge University Press.

Kemp, M.A., & Maxwell, C., 1992. Exploring a budget context for contingent valuation estimates. In: Hausman, J.A. (ed.), *Contingent Valuation: A Critical Assessment.* North Holland, Amsterdam.

Lazo, J.K., Schulze, W.D., McClelland, G.H., & Doyle, J.K., 1992. Can contingent valuation measure non-use values? *American Journal of Agricultural Economics*, (December), 1126-1132.

Mitchell, R.C., & Carson, P.T., 1989. *Using Surveys to Value Public Goods: The Contingent Valuation Method.* Resources for the Future. Washington, DC:

Neill, H.P., Cummings, R.G., Ganderton, P.T., Harrison, C.W., & McGuckin, T., 1994. Hypothetical surveys and real economic commitments, *Land Economics*, **70**, 145-154.

NOAA Workshop, 1994. *NOAA Natural Resource Damage Assessment Workshop Issues.* U.S. National Oceanic and Atmospheric Administration, Washington, D.C.

Ratcliffe, D.A., (ed.), 1977. *A Nature Conservation Review: The Selection of Biological Sites of National Importance to Nature Conservation in Britain. Volume I.* Cambridge University Press, Cambridge.

Slavic, P., Griffin, D., & Tversky, A., 1990. Compatibility effects in judgment and choice. In: Hogarth, P.M. (ed.), *Insights in Decision Making: A Tribute to Hillel J. Einhorn.* University of Chicago Press, Chicago.

Smith, V.K., 1992. On separating defensible benefit transfers from smoke and mirrrors, *Water Resources Research*, **25** (3), 685-694.

Smith, V.L., & Walker, J.M., 1993. Rewards, experience and decision costs in first price auctions, *Economic Inquiry*, **31**, 237.

Stevens, T.H., Echeverria, J., Glass, J.R., Hager, T., & More, T.A., 1991. Measuring the existence value of wildlife: what do CVM estimates really show? *Land Economics*, **67**(4), 390-400.

Tversky, A., Slavic, P.B., & Kahneman D., 1990. The causes of preference reversal, *American Economic Review*, **80**, 204-217.

Whitehead, J.C., Hoban, T.J., & Clifford, W.B., 1995. Measurement issues with iterated, continuous/interval contingent valuation data, *Journal of Environmental Management*, **43**, 129-139.

7 Coastal zone management: the application of networks and databases

T.J. GOODHEAD

Introduction

This paper looks at the background to the potential for using databases and networking by property professionals in the area of coastal and marine resource management. This paper represents part of a wider research project completed in August 1997 and as yet not published.

Marine and coastal policy

Coastal policy

Policy towards the management of the land and water interface has shown little integration in recent years. This is particularly evident in the UK. On the international scale this is at its worst in the third world. Many problems involving coastal management are fundamentally economic with poorer countries having to consider economic factors first, before considering an integrated management system for coastal and marine resources.

There are many uses for the coastal zone and therefore many resources and policies. To minimise conflicts of interest it is important that policies are cross sectorial, but they are not. Many management policies are administrative and vary from region to region. For example, the policy on jet skiers may vary even from estuary to estuary within a particular region. There is a need for an integrated CZM policy at national and international levels. Sea level rises may bring this about. However, global warming is very controversial and disputed by some.

As the world has become more developed, the conflicts in coastal areas and, perhaps to a lesser extent at sea, have become more severe. Coastal protection, large scale clearing, urban development, and aquaculture have led to valuation problems in the areas. These may be specifically identified as:

- urban settlement;
- industrial development;
- waste disposal;
- ports and marine transport;
- land transportation infrastructure;
- water control and supply processes;
- sea fisheries;
- coastal forestry and agriculture;
- extraction industries;
- tourism.

These conflicts and valuation problems would seem, at first site, to generate an intense demand for the skills of property professionals. However, many of the problems involve sensitive environmental issues as shown in Table 7.1 and, as a result, many of the standard valuation techniques used by surveyors may have to be redeveloped or adapted to cope with these issues.

The management of our sea and coasts is creating many problems. Indeed it could be argued that there has been a policy failure in that there has been little policy. However, awareness of the issues in coastal management is so new that it may be unfair to be too critical of policy although there does seem to be a lack of political commitment. Policy has almost developed on an *ad hoc* basis and, as a result, we do not have comprehensive policies or legislation and, indeed, there are often conflicts between the objectives of individual agencies.

Formation of policy towards CZM is difficult since data is often very basic. The methods of collection are not comparable due to the overlapping responsibilities of the agencies and, as a result, mechanisms of intervention are often unsuitable. This can lead to a time lag in terms of intervention.

A concept of 'ecologically sustainable development', developed mainly from UN Conferences (e.g., Agenda 21, 1992) has started to provide a broad framework for developing coastal and marine management systems. The sustainable developments of our resources have become very topical. Definitions of sustainability usually involve a phrase that relates to actions of one generation passing on resources to the next in the condition that they found them, equity being implied between generations. However, the decision often has to be made as to whether we should be selling the 'family silver now'? There are, however, problems of inequality of wealth. The third world may have little chance of developing other resources so are forced to 'ruin' their coastline and seas.

Table 7.1 Critical habitat resources under threat from development

Critical Habitat Resources
Sea grass systems, similar to mangrove and coral
No commercial value and therefore no one fighting to protect it. Trawling, dredging and sedimentation creating problems. However, tourism perhaps gives value to these systems.
Coral reefs
Fishing creates problems as high seas fish species depend on reefs for their juvenile stages. Tourism can create a major impact and reefs are thought to be dying. Enemies are tourism, fishing and dynamite fishing.
Sandy beach systems
Dynamic environments that have conflicts of tourism and mining.
Tidal flats
Very important feeding areas. There are conflicts with recreation and development.
Lagoons and estuaries
Very vulnerable ecosystems. There are conflicts with commerce, property development and recreation.
All areas: pollution
Domestic sewage and public health.

There is a need to discern between renewable and non-renewable resources. It is difficult to estimate the world-wide stock of resources other than for some mammals. New sources of energy may well come on line. There is still a climate of rapid technological development in the UK. It is very important to consider economic sustainability and product life cycles. Population control may be a key factor in the management of coastal and marine resources but this is rarely considered.

Within this background of fragmented policy towards coastal and marine resource management at an international level, there has developed a similarly fragmented policy towards coastal and marine zones in the UK. The trump card for the surveyor is perhaps that valuation and property management is likely to become of increasing importance in the management decision making process. Given the scale of resources that can be considered under the general theme of marine resource management, there would appear to be tremendous scope for new business and activity for property professionals.

The United Kingdom policy

Above the low water mark. In the UK, government management of the 'Property Life Cycle': land and sea, measurement, cultivation, planning, construction, ageing, management, investment, development, refurbishment, and demolition is largely a function of a cascaded planning system that operates down to the low water mark. A review of this process can be found in Chapter 8 of this book and by Taussik (1996). The sectorial approach to managing below low water mark is so complex that there are issues that affect all property professionals.

Below the low water mark. For this sector, an approach to management exists for:

- construction projects outside harbours;
- harbour construction and construction within harbours;
- marine dredged minerals;
- navigation and other dredging;
- disposal at sea;
- pontoons and other moorings;
- marine fish and shellfish farms;
- oil and gas exploration and production;
- submarine pipelines;
- submarine cables.

This sectorial approach to the management of development below the low water mark or territorial sea has resulted in a complex web of legislation, agencies and professionals. In terms of this research project, the sectorial approach to management has resulted in a raft of interests and professional organisations becoming involved in Marine Resource Management. It is critical that the property professionals promote its members own interests and network with these organisations since competition for business is now very competitive.

This approach to the management of the UK maritime resources is, at first sight, bewildering. However, it could be argued that the operator in one sector does not need to know a great deal about the other sectors. As conflicts arise there appears to be a need for trained professionals who can both assign values and provide solutions to these conflicting interests. This is potentially a great growth area for chartered surveyors generating business in this area.

Since the UNCLOS agreements of the 1980s have slowly gained acceptance, there has been an awareness that since the world's water space has been broadly divided up and the technology for exploitation has been developed, that management issues will now come to the fore in the following three zones:

- the *Territorial Sea*, defined as the area of up to 12 miles offshore from a coastal state;
- the *Exclusive Economic Zone*, defined as the area of up to 200 miles offshore from the coastal state;
- the *Deep Sea*, an area of water mass that is at least 200 miles from the coast.

The acceptance of the Exclusive Economic Zone by coastal states will develop opportunities for RICS surveyors. The Department of Trade and Industry has established a BREEZE project (British Exclusive Economic Zone Executive) to market British skills in the EEZ overseas. Valuation of resources within the EEZ would seem to be fundamental to successful management.

Transport

Freedom of navigation and innocent passage is a concept that goes back to the Victorian era and the rights of vessels at sea have been subject to a number of international agreements that follow from an era when Great Britain was a major maritime power.

135

Whilst the Department of Transport has the right to control development in tidal waters to safeguard navigation, the Coast Protection Act (1949) and the COLREGS (international collision regulations developed from English domestic rules and later annexed to multi national treaties) effectively govern the right to and innocent passage. These have been involved initially from the SOLAS 1929, 1948 and 1960 agreements and subsequently the Convention on the International Regulations for Preventing Collisions at Sea (1972). The COLREGS Convention relates to safety and not efficiency. The COLREGS are enforced by threat of discipline from the state that has provided the flag, or in whose territorial waters the incident has occurred. Civil cases are held in the country where the underwriters reside, normally either the UK or the USA. These regulations can be backed by the International Maritime Organisation (IMO) clarifications.

However, not all countries are confident about the IMO, which is the first international body devoted exclusively to maritime matters. Its governing body is the Assembly, which meets once every two years. Its consists of 137 Member States and two associated Members. From the very beginning, the improvement of marine safety and the prevention of marine pollution have been the IMO's most important objectives. The 1972 COLREGS came into force in 1977 and have now been amended four times. New developments such as hovercrafts and high speed ferries will inevitably result in further changes such as the 1982 Law of the Sea Convention (LOSC). The 1992-93 Braer and Agean Sea disasters led to the Particularly Sensitive Sea Areas concept (PSSAS) resulting in improved routing and ship reporting. it is likely that in future, the relative independence of maritime agencies from strategic marine resource management decision making will effectively end, in view of the heightening awareness of global environmental issues. This will open up even further opportunities for surveyors who have the capability of developing integrated management systems linking commercial shipping and the built and natural environment. The surveyors skills in estate management are transferable to the coastal and marine zone, evidence of this is the use of firms of surveyors by the Crown Estates to manage the foreshore. The transfer of these skills to the management of marine resources including, perhaps the management of transport routes, could lead to a substantial growth in new business.

The key networking organisations

The United Kingdom

Within the UK, a number of marine and coastal networking forums exist, e.g., Arforid, CoastNet, the National Coast and Estuaries Advisory Group, Wildlife and Countryside Link – Marine Groups, Coastal Resiarch and Management Group of the JNCC, European Union for Coastal Conservation, Eurocoast. The most notable of these is probably CoastNet, a coastal heritage network. Membership provides access to a network of coastal managers, a bulletin, training, research reports, a library and a number of other services. Established in 1996 by field staff and managers, it has provided an open membership and is supported by national agencies. It has recently produced a UK Coastal Management Directory (CoastNet, 1997) to aid communications, networking and exchanges between those involved with the UK's coastal and marine environment. The potential consumer base for this networking forum, and the others, has been identified by the Centre for Coastal Zone Management (Pickering, 1994) as:

- Heritage Coast officers;
- fisherman and representatives;
- mineral extraction companies;
- landowners;
- recreation and tourism groups;
- local authorities;
- academics and students;
- the user community;
- conservation groups;
- harbour authorities.

There are many other organisations fulfilling a networking role in the UK. For example, the British Marine Industries Federation has a yearbook listing 3,500 members and the Royal Yachting Association has a membership of approximately 80,000 to name but two.

In terms of professional organisations being involved with Marine Resource Management, there appears to be six, as identified by Pickering (1994):

- Institute of Ecology and Environmental Management;
- The Royal Town Planning Institute;
- The Institute of Leisure and Amenity Management;
- Institution of Civil Engineers;
- Institution of Water and Environmental Management;
- The Royal Institution of Chartered Surveyors.

To this list should be added the professional organisations representing the mariners and their supporting professions. The small number of organisations mentioned above is part of a much larger group of organisations that are involved with marine resource management. The UK Coastal Management Directory (1997), lists more than 1,300 in the UK alone and divides them into:

- local authority staff;
- coastal site managers;
- coastal zone management project staff;
- government agencies;
- voluntary bodies;
- regional Ministry of Agriculture, Fisheries and Food offices and fisheries;
- sea fisheries committee officers;
- port and harbour masters;
- industry, fisheries and shipping interests;
- sport, recreation and tourism;
- farming and land owning bodies;
- coastal networks;
- research organisations.

At a European level, Eurocoast represents 11 national associations (Eurocoast, 1997) and has its secretariat at Cardiff University. Although relatively small in size with only 150 individual members and eight corporate members, it is proposed that it be merged with CoastNet.

Opportunities

Within, for example, the Royal Institution of Chartered Surveyors (RICS) there are opportunities for linkages to be made between members of existing divisions. There are existing mechanisms for this, notably through their regular publications. Outside of the RICS, as networking organisations have evolved, there is a danger for the RICS that those organisations controlling the formation of the networking database will be in a position of considerable commercial power. The valuable role of the Chartered Surveyor in Marine Resource Management may be simply swamped by the vast number of organisations in these databases being produced by networking organisations.

Strategy for a property profession in networking and generating databases

This is based on the research undertaken in July and August, 1997 at the University of Portsmouth (Goodhead, 1997):

- in-house magazines are a vital tool in networking throughout the property professions but this assumes that members will contribute towards it;
- the need for an up-to-date brochure was stated in the majority of responses;
- the production of publications seemed to be important in networking;
- the surveyors questioned had an amazing diversity of knowledge, having read widely and attended a wide variety of conferences. There must be scope for sharing this knowledge;
- closer links could be made with Government initiatives, such as the Department of Trade and Industry's BREEZE project;
- there is scope for creating interdisciplinary teams across the property professions relating to marine resource management;
- property professionals could join up with a specialised networking group, such as CoastNet, rather than creating its own;
- social meetings were held in very low regard, although the idea of a single networking workshop was held in high regard;
- a limited number of surveyors had access to a database but a substantial number of respondents said that they would be prepared to contribute to one;

- the concept of providing a feedback report of conferences attended was supported by approximately two thirds of those who answered the questionnaire.

Opportunities and threats

It is clear that there are opportunities for the surveyor and property professionals in this area. The future use of our coats is likely to require further development and exploitation. The management of this development will require some complex valuations as cost benefit analysis is applied by the decision-makers.

The threat to the property professions is that many other professionals bodies are moving into the marine resource management arena and, for example, the surveyor will have to compete hard for a share of the market for its members. The civil engineers, in particular, through the development of coastal defences are by default, developing skills in valuation and property management. In the UK, the planners are restricted, by statute in their activities, to the low water mark but the creation of various Coastal Forums has enabled them to develop an offshore expertise.

Strategy

In the light of the above comments, the following recommendations are made regarding the development of a strategy:

- develop a concise marine resource management brochure;
- develop a rapporteur system to link with other Divisions of the property profession;
- assign one marine resource management member to 'cover' and report back on other marine and coastal organisations, such as Eurocoast, CoastNet, etc.;
- develop a database similar to the Eurocoast model. This database would provide a directory of experts in marine and coastal management that could be distributed for marketing purposes. This could form the the basis of a directory of experts;
- send key people to conferences, paid for by the profession and as a condition provide a written report and feed back to the annual conference;

- develop a centrally issued briefing sheet or information page in the profession's magazine;
- provide an annual student social meeting;
- hold annual conferences but change the emphasis towards a national meeting point, providing feed back from annual conferences that members have attended, listing key people met and major institutional contacts.

Conculsion

There is great scope for the property professional generating business in the coastal and marine zones. Due to the complex nature of policy, success in this area may be a function of the ability to liase and network with other professionals to form multi-disciplinary teams. As a consequence, effective networking and marketing is vital.

References

Agenda 21, 1992. *Chapter 17. Protection of the Oceans, All Kinds of Seas, Including Enclosed and Semi-Enclosed Seas and Coastal Area and the Protection, Rational Use and Development of their Living Resources,* United Nations Conference on the Environment and Development, Rio de Janeiro.

Centre for Coastal Zone Management, 1994. *The UK Heritage Forum,* University of Portsmouth.

CoastNet/NCEAG UK, 1997. *Coastal Management Directory.* CoastNet.

Eurocoast, 1997. *Membership Directory.*

Goodhead, T. & Johnson, D., 1996. *Coastal Recreation Management,* SPON.

Goodhead, T., 1997. *Networking and Databases,* unpublished, University of Portsmouth.

Pickering, H., 1994. *Professional Affiliation Options,* Centre for Coastal Zone Management, University of Portsmouth.

Taussik, J., 1996. Planning and the provision of marine recreation facilities. In: Goodhead, T., & Johnson, D., 1996. *Coastal Recreation Management,* SPON.

8 The contribution of Town and Country Planning to the resolution of conflict in the coastal zone

J. TAUSSIK

Introduction

The coastal zone represents the interface between land and sea. It extends to include those areas of land and sea where there is interaction and comprises four elements: the land, the seabed, the intertidal zone and the water. The juxtaposition of land and water elements presents a range of opportunities for human exploitation including, for example: communication and navigation; water based recreation; fisheries and aquaculture; the disposal of waste materials. Additionally, the coastal zone provides suitable locations for a wide range of land uses, including agriculture, manufacturing industry and tourism and, with its economic and aesthetic attractions, is much favoured for residential development. However, the juxtapositioning of land and sea also brings threats and risks to human activity, for example, from erosion, tidal flooding or coastal instability. Also, the very richness and variety of coastal environments may pose a constraint on human activity because of the fragility or rarity of a landform and/or habitat.

It is clear from the few examples quoted that not all of these uses of the coast are likely to co-exist in a geographical area without problem; indeed, conflict is likely between the wide ranging interests in the coastal zone. These are summarised on Figure 8.1 and it is suggested that conflict extends not only between almost all of the sectors (for example: between waste disposal/pollution control and fisheries/harvesting/aquaculture or recreation or nature conservation; or between shipping/navigation/communication and recreation) but, also, within the sectors (for example: between different sectors of recreation; or the effects of engineering works on coastal processes over an extensive area). Neither are the conflicts limited in their geographical location.

The moving nature of water and the water mass means that today's local problem becomes tomorrow's regional/national problem and an international problem the day after.

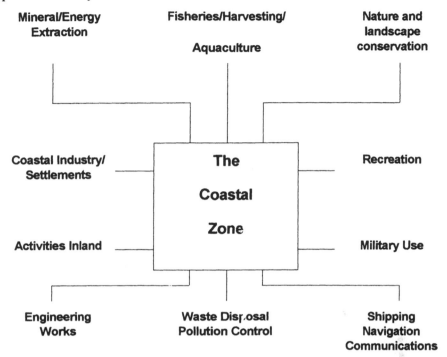

Figure 8.1 Major uses in the coastal zone

While certain problems have been recognised for some decades, as the level of human exploitation of the coastal zone has increased, these conflicts have become more apparent. It is estimated that human dependence on coastal resources has increased so that, now, over 60% of the world's population live in the coastal zone (Reid, 1997). This is expected to increase further. In fact, it has been estimated that there will be more people living in the coastal zone in the year 2025 than there are currently living in the whole world now. At that time, nineteen of the expected twenty-two megacities with populations over 10 million persons will be coastal (Reid, 1997). This suggests that the conflicts and problems of the coastal zone will increase in scale and complexity and reinforces the importance of accepting the principles of coastal zone management. OECD (1993) implicitly emphasises the need for management when it identifies the characteristics of coastal resources as that:

- they behave as systems and should be managed as systems to provide sustainable outputs;
- they interact amongst each other. Sustainable management must take these interactions into account;
- they have multiple uses and they can deliver simultaneously many sustainable outputs;
- complementarities and conflicts exist between and among various uses. To achieve ecologically sustainable development these conflicts have to be resolved and trade-offs made;
- coastal resources supply private goods as well as public goods from which nobody can be excluded. Non-exclusion can lead to overuse of a resource and threaten its sustainability; and
- they can be supplied by the private or the public sectors and sometimes are supplied by both. These operations need to be co-ordinated.

Coastal zone management

Coastal zone management has been receiving increasing levels of attention internationally, particularly since its inclusion in Chapter 17 of Agenda 21, adopted at the United Nations Conference on Environment and Development (UNCED) (The Earth Summit) at Rio de Janeiro in June 1992. Developed from management approaches to deal with rural resources, coastal zone management is an umbrella term which encompasses many methods and modes but with the underlying objective of moving towards a more sustainable use of coastal resources.

Although there are many other definitions (Gubbay, 1990), coastal zone management may be defined as

> a dynamic process in which a co-ordinated strategy is developed and implemented for the allocation of environmental, socio-cultural and institutional resources to achieve the conservation and sustainable multiple use of the coastal zone (CAMPNET, 1989).

It is also known as integrated coastal zone management, to reinforce its strategic or overarching character (geographically, institutionally and/or by activity), or, simply, as coastal management.

The aims of coastal zone management may be considered to be:

- to promote sustainable use;
- to balance demand for coastal zone resources;
- to resolve conflicts of use;
- to promote environmentally sensitive use of the coastal zone; and
- to promote strategic planning for coasts (Gubbay, 1990).

Other authors explicitly emphasise the role of coastal zone management in resolving conflicts of use:

> Coastal zone management typically is concerned with resolving conflicts among many coastal uses and determining the most appropriate use of coastal resources (Sorensen *et al.,* 1984, cited in Gubbay, 1990).

The review of Cicin-Sain *et al.,* (1995) concerning progress achieved in the implementation of integrated coastal zone management and the further development of related guidance prepared by international agencies suggests that this is the favoured approach for resolving the conflicts of interest in the coastal zone. However, this consensus of approach and of overall objective masks a wide range of methods and modes by which coastal zone management is operated, organised and undertaken, as evidenced by papers presented at the increasing numbers of conferences dedicated to coastal zone management. This reflects the various contexts in which coastal zone management operates and includes varying geographical scale, organisational scope, level of horizontal and/or vertical integration, the time horizon and the level of community involvement.

Coastal zone management in the U.K.

As in other European countries, there is no statutory framework for coastal zone management in the U.K. While the Government (DOE, 1992a) accepted the House of Commons Environment Committee's (1992b) recommendations supporting the need for coastal zone management, it did not consider it was feasible to consolidate the legislation affecting the coastal zone or to establish a national coastal zone unit to adopt a national overview of coastal zone policy. It did not believe that there was a widespread duplication of responsibilities or poor co-ordination in coastal management. The opportunity to create an integrated statutory system for the management of coastal resources in the UK, therefore, was not taken and these resources continue to be managed on a sectoral basis with a large number of organisations being involved (DOE, 1993). Each has different powers and operates within different geographical boundaries (see Figure 8.2). For example, development of land is controlled, primarily, through the town and country planning system while the development of the seabed is regulated by the Crown Estate, through leasing and licensing arrangements. Coastal defence is split between coastal protection (protection of land from erosion) and sea defence (defence of land from flooding), each administered by different organisations. Elements of water quality, discharges to water and disposal of waste at sea, are dealt with by different administrations. Remembering that institutional systems vary between England, Wales, Scotland and Northern Ireland, an extremely complex and confusing institutional environment exists for those operating in the coastal zone of the U.K.

The advantages of integrated approaches are increasingly recognised but initiatives in the U.K. can only be undertaken through non-statutory, voluntary means. Current initiatives include the preparation of:

- *Coastal zone management plans.* These are intended as strategic documents incorporating land and water areas and creating a framework in which decisions about the coastal zone can be made on an integrated, rather than a sectoral, basis. Existing coastal zone management plans are limited in number though some strategic documents have been prepared, usually with local authorities as the leading agency in their preparation.
- *Estuary and harbour management plans.* These plans relate to more restricted areas and cover partially contained water areas and surrounding, associated land. They provide an overview of a wide range of estuary users and establish policies for the sustainable use of such areas. A substantial number of such plans are under preparation by a range of organisations.
- *Catchment management plans.* Catchment management plans were prepared by the National Rivers Authority (NRA). They related to the water collection areas of river basins and provided a strategy for the use of the area to protect and improve water quality. They cover large land areas with limited coastal extent but can include estuaries

and harbours. They have now been replaced by Local Environment Agency Plans (LEAPS), prepared by the Environment Agency as successor to NRA and covering a wider brief.

- *Shoreline management plans.* These plans extend over a relatively narrow belt of land at the shoreline and relate to the coastal cells in which coastal processes are relatively self-contained. Their purpose is to set out the strategy for the protection of coastal land from natural processes associated with the coast.
- *Other plans.* These include heritage coast management plans, management plans prepared for the use and conservation of areas of coast with landscape and/or wildlife value or where there is heavy recreation demand needing to be management. This heading also includes non-statutory plans prepared in support of policies for the regeneration of developed areas such as docks and resorts.

All of these initiatives are non-statutory. They all involve the co-operation and co-ordination of a range of sectoral interests to secure jointly agreed objectives. Such voluntary approaches offer a range of advantages. They require exchange of information, enhance understanding of multi-sectoral problems and involve sectoral interests in the implementation of mutually agreed policies. Where support for these policies remains consistent and the policies are implemented, the voluntary approach has much to recommend it. However, their voluntary nature is also a potential weakness. Where commitment to agreed policies is weak or circumstances change, major players may abandon their commitment to jointly agreed policies in order to achieve narrower, sectorally defined objectives. This undermines the agreed plans. Voluntary approaches to coastal zone management, therefore, are prone to uncertainty even where statutory systems for regulating the use of coastal resources exist, as in the U.K.

The role of coastal zone management in reducing conflict in the use of coastal resources was outlined above. The plans and strategies described are only as good as their implementation. Plans with explicit proposals, clear policy frameworks and beautiful presentation but which rest on official shelves without effecting change are of no value at all. In fact, the existence of an unimplemented plan will be a greater hindrance to the acceptance of coastal zone management ideas than having no plan as credence previously given to the management activity will be undermined. It is clear, therefore, that coastal zone management will move towards assisting in the resolution of conflict only where the plans and strategies that are prepared are implemented.

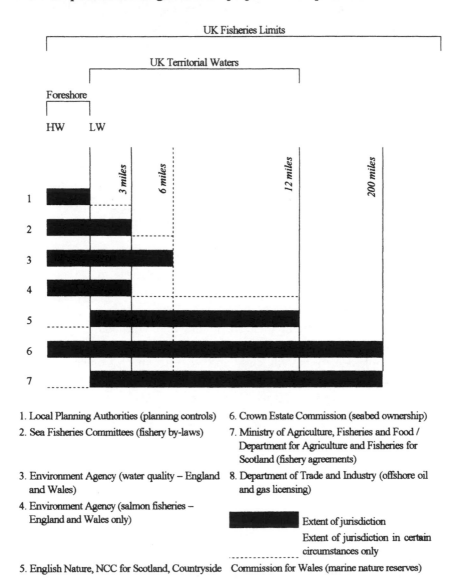

1. Local Planning Authorities (planning controls)
2. Sea Fisheries Committees (fishery by-laws)
3. Environment Agency (water quality – England and Wales)
4. Environment Agency (salmon fisheries – England and Wales only)
5. English Nature, NCC for Scotland, Countryside Commission for Wales (marine nature reserves)
6. Crown Estate Commission (seabed ownership)
7. Ministry of Agriculture, Fisheries and Food / Department for Agriculture and Fisheries for Scotland (fishery agreements)
8. Department of Trade and Industry (offshore oil and gas licensing)

Extent of jurisdiction

Extent of jurisdiction in certain circumstances only

Figure 8.2 Sectoral management of coastal resources (after Gubbay)

The voluntary nature of the plans, with the associated implementational uncertainty is, therefore, a limitation of coastal zone management in UK and, indeed, in any other country employing a similarly voluntary approach. It is with this limitation in mind that it is proposed that the town and country planning system can assist in the resolution of conflict in the coastal zone.

The town and country planning system

Like other European countries, the U.K. has a well established, statutory system of town and country planning. There is wide recognition of its role in the allocation of, and resolution of conflict over, land and property resources (Healey, 1992). Planning systems tend to incorporate three elements:

- Forward planning. In England and Wales this involves development plans at strategic and local levels;
- Control of development. This ensures that development undertaken conforms to proposals and policies in plans; and
- Compulsory purchase. Currently, little used in the U.K., this operates to overcome market failures.

In the U.K., the operation of the system is administered by local planning authorities but is overseen by central Government, which establishes the overall policy framework and retains final powers of decision making.

The term 'development plan' is an umbrella term for a range of plans setting out the policies and proposals for the use and development of land and property in a particular area. The whole of England and Wales is to be covered by development plans comprising both strategic and detailed levels.

Unitary development plans (UDPs) are prepared by unitary authorities, London Boroughs and metropolitan districts. They are divided into two sections: Part 1 deals with strategic matters; and Part 2 covers matters of detail.

Structure plans are strategic plans, each covering the administrative area of a county. They are non-specific in their content and include policies of a broad and general nature.

Local plans provide the detail to support structure plans and are generally prepared for district areas through local plans for mineral operations and for waste cover whole county areas.

In both the administration of development control and compulsory purchase, the development plans in England and Wales carry great weight. They will have undergone public scrutiny and, unless there are overriding considerations, which dictate otherwise, development decisions must be based on them. Therefore, they are fundamental to the operation of the system and carry great influence over what can and cannot be done with land and property. They do not, however, operate in isolation but within the context of Government established, national policies, including for the coast, to create a cascade of planning policy for a particular area.

Integrating coastal management and town and country planning

Policies in coastal management plans cover a wide range of subjects: fisheries; water quality; navigation and communication; recreation; conservation of man made and natural environments; coastal defence; and so on. Very many of the subjects incorporate a land use/development dimension. For example, storage, processing and marketing facilities are required in association with the fishing industry. Launch points, slipways, toilets, moorings and car parking are integral aspects of recreation activity. The control of development which may pollute ground waters, which may threaten wildlife resources or unspoilt views, or which may have severe implications for tidal regimes or be at risk of coastal flooding or erosion is undertaken through the planning system rather than through other regulatory mechanisms. Other objectives, like improving public access to the coast or enhancing coastal environments, may be achieved through the development process. The planning system, therefore, can be seen to be integral to the processes by which some coastal management objectives are achieved and plans implemented. In this, development plans offer opportunities for the incorporation of policy matters from coastal management plans, both at the strategic and the detailed level, into a statutory framework (see Figure 8.3).

It is clear, therefore, that there should be a close relationship between coastal management plans and development plans. This is illustrated in Figure 8.4. Firstly, as development plans are well established and coastal management planning is a new activity, development plans set the context for and inform coastal management plan preparation on policies and proposals for land use and development in the coastal zone. However, this process of informing should be two-way. Coastal management plans, with their greater emphasis on all four of the elements of the coastal zone, can inform development plans about the non-land-based issues. They also offer opportunities to reinforce aspects of development plans by providing policies complementary to, but beyond the scope of development plan policy. This could include, for example, policies related to the use of water areas for recreation or on public access to the coast.

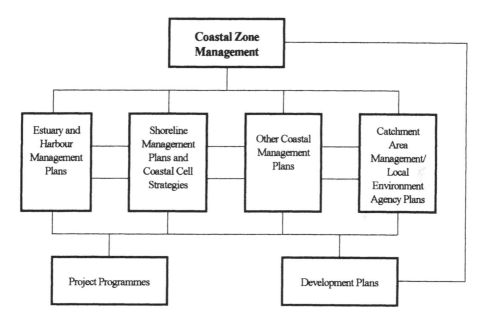

Figure 8.3 The relationship between different coastal plans

It is suggested that the statutory nature and importance of development plans means that they offer an opportunity to enhance the non-statutory status of coastal zone management plan policy in the context where the voluntary nature of these plans creates a level of uncertainty over implementation. Such policies then become an integral part of the planning system and influence development related decision making. There are, however, major limitations in the extent to which development plans can pursue policies and proposals for the coast. First, development plans can include only matters related to the use and development of land and property, that is, to planning matters. Matters like water quality are not, directly, planning matters although the affect of development on water quality is a planning matter. The statutory planning system, therefore, is limited in its scope in contributing towards resource management. A second problem is that the planning system in the U.K. operates normally only to the low water mark and does not incorporate the seabed or the water mass. Central government guidance for England and Wales (DOE/WO, 1992) suggests that, although development plans cannot include

151

policies for the land mass and water area, planning authorities should consider the effects off-shore of development on land. This is not happening. Planning authorities, while giving higher priority to coastal matters than in the past, demonstrate a 'land by the sea' approach to policy (Taussik, 1996a). This appears to reflect their land based tradition and limited knowledge of the coastal zone and emphasises the need for interaction between the two types of plan.

It was suggested by the House of Commons Environment Committee (DOE. 1992b), but not accepted by the Government (DOE. 1992a), that the British planning system should be extended off-shore. This would broaden planning's contribution to coastal management and could allow some rationalisation of the regulation of the use and development of coastal resources. However, experience in Sweden suggests that changing the institutional system alone will not achieve environmental objectives (Taussik, 1996b).

Like Britain, Sweden has a well-established planning system. High priority is given to the quality of the coastal environment. Unlike Britain, the Swedish planning system is required to plan for land *and* water areas (including coastal waters). Unlike Britain, local government administration extends seawards to the limit of the territorial sea. This suggests that the Swedish planning system, and forward planning in particular, should play a major role in the management of coastal resources. Experience shows that this has not happened. The planning system is as land bound as in England and Wales. The necessary integration of planning with other activities has not occurred. Swedish experience suggests that other pre-requisites, notably information and awareness and political support, as well as the existence of appropriate institutional frameworks, are necessary before environmental objectives will be achieved.

Conclusions

The British Government has confirmed its commitment to a coastal management system based on a sectoral approach, with different management systems operating in different parts of the coastal zone. None the less, there is increasingly widespread appreciation of the benefits of coastal zone management. This underlies the preparation of increasing numbers of voluntarily prepared coastal management plans. These plans tend to be very new and, to date, little has been done to incorporate appropriate elements of these plans in development plans. However, there is much benefit from incorporating such material in development plans with their statutory status. Further, coastal management plans can inform development plans so that development plan policies become truly coastal, instead of being policies for land that happens to be by the sea. However, for the British planning system to take a properly coastal perspective may require changing the institutional framework to extend planning seawards. To be successful, such a change must be accompanied by increased knowledge and awareness and enhanced political support.

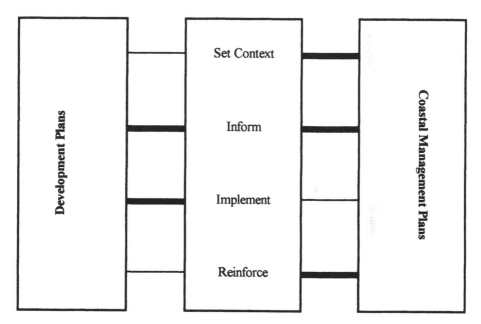

Figure 8.4 The integration between coastal plans

This paper has concentrated on experience in Britain. With its well-established planning system and more recent acceptance of coastal zone management operated through voluntary means, it is typical of many developed countries, particularly in Europe. The lessons from and suggestions about Britain can, therefore, be more generally applied at a European level and beyond.

References

CAMPNET, 1989. *The status of Integrated Coastal Zone Management: A Global Assessment.* Preliminary summary report of a workshop convened at Charleston, South Carolina, July 4-9. 1989.

Cicin-Sain, B., Knecht, R.W., & Fisk, G.W., 1995. Growth in capacity for integrated coastal management since UNCED: an international perspective. *Ocean and Coastal Management*, **29**, Nos 1-3, 93-123.

Department of the Environment, 1992a. *Coastal Zone Protection and Planning: The Government's Response to the Second Report from the House of Commons Select Committee on the Environment.* HMSO, London.

Department of the Environment, 1992b. *Coastal Planning and Management: A Review.* HMSO, London.

Department of the Environment, 1993. *Coastal Planning and Management: A Review.* HMSO, London.

Department of the Environment and Welsh Office, 1992. *Planning Policy Guidance: Coastal Planning.* PPG20. HMSO, London.

Gubbay, S., 1990. *A Future for the Coast? Proposal for a U.K. Coastal Zone Management Plan.* Marine Conservation Society, Ross-on-Wye.

Healey, P., 1992. The Reorganisation of State and Market in Planning, *Urban Studies*, **29**, 411-434.

Organisation for Economic Co-operation and Development, 1993. *Coastal Zone Management. Integrated Policies.* OECD, Paris.

Reid, W.V., 1997. *Human Health and Coastal Management: Social and Demographic Driving Forces.* Paper presented at Coastal Zone 97, Boston, Massachusetts, 19-25, July 1997.

Taussik, J., 1996a. Development plans and the coastal zone, *Town Planning Review*, **67**, 397-420 (also available as WP 17 of Working Papers in Coastal Zone Management. University of Portsmouth, Portsmouth).

Taussik, J., 1996b. A comparison of coastal planning: England and Wales and Sweden. In: Y. Rydin (ed.). *The Environmental Impact of Land and Property Management.* Wiley, London.

9 Partnership in planning and management of the Solent

A. INDER

Introduction

The Solent is centrally positioned on the south coast of England. It is the area of relatively sheltered inshore waters between Hampshire and the Isle of Wight including Southampton Water, several estuaries and Portsmouth, Langstone and Chichester Harbours.

It is one of the most important coastal zones in England, for a number of reasons. There are three Areas of Outstanding Natural Beauty (South Hampshire AONB, Chichester Harbour AONB and the Isle of Wight AONB) and two stretches of Heritage Coast (Hamstead Heritage Coast and Tennyson Heritage Coast) within the Solent. Its extensive, biologically rich, inter-tidal areas support over 100,000 waders and 40,000 wildfowl in the winter, making the Solent one of the top five coastal areas for bird populations in the UK and of international importance for wildlife. It is a very busy waterway; Portsmouth is the country's second busiest continental ferry port, and Southampton is the second most important port for ocean liners and container ships. Portsmouth is the UK's premier naval base. The Solent is the busiest sailing ground in Western Europe and possibly in the world, and is also important for other forms of water recreation such as windsurfing. The area's vital role in the defence of the realm over the centuries has resulted in the nation's most extensive and important coastal defence fortifications, a rich historic heritage which attracts millions of tourists.

Therefore, this unique and special coastal area is the focus for some important uses and activities, and the arena for the inter-play of competing and potentially conflicting interests.

Table 9.1 Organisations involved in the planning and management of the Solent

A. Key regional or local government authorities	
Authority	**Responsibilities and powers**
Local Authorities (County Councils; City, Borough and District Councils; Unitary Councils)	Wide-ranging responsibilities including preparation of statutory plans (e.g. structure and local plans); development control; environmental health; leisure and recreation; coast protection; environment/heritage conservation. Important landowners. Administrative jurisdiction normally ends at low water mark.
Harbour Authorities	Wide-ranging powers within their harbour areas, especially for harbour conservancy and navigational safety. Some statutory bodies (e.g. Chichester Harbour) have responsibility for recreation management and landscape/nature conservation. Some harbours are under local authority control (e.g. Langstone; River Hamble).
SERPLAN (The London and South East Regional Planning Conference)	Standing Conference of local planning authorities. SERPLAN has a regional strategy, which includes coastal policies, and has produced Coastal Planning Guidelines.
SCOPAC (Standing Conference on Problems Associated with the Coastline)	A non-statutory body comprising local authorities and other bodies with statutory responsibilities for coastal defence and related matters. Provides a co-ordinated approach to coastal defence works, and encourages the preparation of shoreline management plans. Commissions important research projects and facilitates exchange of information. Extends from West Dorset to Arun District (West Sussex).
Solent Water Quality Conference (Standing Conference)	A non-statutory body comprising local authorities, to assist in protecting and improving the quality of recreational waters. Encourages a co-ordinated approach to monitoring, and to tackling water quality issues.

B. Key central government departments and other statutory national bodies. Responsibilities and powers

Authority	Responsibilities and powers
Countryside Commission	Designates or identifies areas which have particular scenic value and which need protection from adverse development. Areas of Outstanding Natural Beauty are statutory, and Heritage Coasts are defined in development plans by local planning authorities in consultation with the Commission.
Crown Estate	A landed agent and major landowner. Neither government property nor the private estate of the Sovereign. The Commissioners have a duty to maintain and enhance the capital value of the estate and the income obtained from it. Owns much of the foreshore and tidal river beds between mean high water and mean low water, and the territorial seabed in the Solent. Provides aggregate dredging licences, and leases seabed for the management of moorings etc.
Department of the Environment, Transports and the Regions (DETR)	Government department responsible for designating or approving the designation of national or international nature conservation sites and areas, in consultation with English Nature (or other country agencies) and the Joint Nature Conservation Committee. The government sponsor for the Environment Agency and English Nature. Committed to prepare national strategies and programmes for the conservation and sustainable use of UK biodiversity. Has produced Policy Guidelines for the Coast, and runs the Coastal Forum.
English Nature (EN)	Government's main statutory advisor on nature conservation. Responsible under the Wildlife and Countryside Act 1981 for the establishment of National Nature Reserves, notification of Sites of Special Scientific Interest and selection of Marine Nature Reserves. Advises on the selection and management of Special Protection Areas, Ramsar sites and Special Areas of Conservation. Promotes the conservation of England's wildlife and natural features.
The Environment Agency	A non-statutory public body with statutory duties and powers relating to flood defence, water resources, pollution control, waste regulation, fisheries, recreation, conservation and navigation.

157

Ministry of Agriculture, Fisheries and Food (MAFF)	Overall responsibility for policy and grant aiding coastal defence. Controls dumping at sea. Also, administers legislation controlling aspects of fisheries protection, for example Fisheries Orders. Enacts bylaws at the request of Sea Fisheries Committees to control fisheries within the 6 nautical mile limit.
Ministry of Defence (MoD)	Owns many coastal sites, from largely natural areas to dockyards and buildings. Controls a large sea area of the Solent through the Queen's Harbour Master. Crown immunity confers MoD with freedom from certain statutory requirements imposed on other coastal land users.
Southern Sea Fisheries Committee, Sussex Sea Fisheries Committee	Responsible for managing and policing the inshore fisheries out to 6 nautical miles. Bylaw-making powers through MAFF to regulate, protect and develop fisheries. May make bylaws solely for environmental purposes, including protection of the marine and coastal environment (in order to implement the EC Habitats Directive) and protection of physical and archaeological features.
The Sports Council	Established by Royal Charter in 1972, the Sports Council aims to increase participation in sport and physical recreation, increase sports facilities, raise performance standards and provide information.
HM Coastguard	Responsible for the initiation and co-ordination of civil maritime search and rescue operations.
Marine Safety Agency	An executive agency of the Department of Transport. responsible for implementation of the government's strategy for marine safety and prevention of pollution from ships.

C. Commercial Organisations

Commercial ports and shipping	Associated British Ports (Southampton) Portsmouth Commercial Docks Board Southampton Shipowners Association British International Freight Association
Other commercial organisations	British Petroleum Esso British Marine Aggregates Producers Association British Marine Industries Federation Southern Water Services

D. Environmental Organisations (Voluntary)
Hampshire & Isle of Wight Wildlife Trust Hampshire & Wight Trust for Maritime Archaeology Royal Society for the Protection of Birds Solent Protection Society Council for the Protection of Rural England

E. Recreation Organisations (Voluntary)
Royal Yachting Association Solent Cruising Racing Association

F. Academic Institutions
University of Portsmouth University of Southampton Southampton Institute of Higher Education

As for any other coastal zone in the UK, the basic organisational structure for planning and management in the Solent is highly fragmented. Many types of statutory and non-statutory authorities and agencies have a role to play, each with its own powers, objectives and priorities, and area of jurisdiction (see Table 9.1). Even within groupings of authorities there are significant variations, for example in the powers and responsibilities of the eight harbour authorities in the Solent (and the main shipping channel in the western Solent is outside the jurisdiction of any harbour authority).

The planning framework is based largely on the statutory Town and Country Planning system. Whilst it is well established and has many strengths, it is restricted to land use and development and is unable to tackle management issues. And, most importantly from the coastal point of view, jurisdiction normally ends at the low water mark, although in the Solent there are significant exceptions.

The pressures on the area are increasing, too. Since 1951 the population of the Solent area has increased from 716,000 to more than 960,000, and boat numbers increased from 23,000 to nearly 33,000 between 1978 and 1991. The volume of trade handled by the ports at both Southampton and Portsmouth has increased enormously in recent years.

These pressures have increasingly exposed the inadequacies of the fragmented system of administration and have led to calls for a more integrated system of planning and management, and a more long-term, strategic approach.

The first moves in the right direction came in the late 1980s, initially through the grouping of like minded organisations on a sectoral basis.

For example, in 1986, at the instigation of the Isle of Wight County Council, the Standing Conference on Problems Associated with the Coastline (SCOPAC) was formed to bring together the local authorities (responsible for coast protection) and the National Rivers Authority' (responsible for sea defence) along the central south coast from Weymouth to Worthing, including the Solent. This enabled a more strategic and co-ordinated approach to be taken to coastal defence, and environmental considerations have been integrated better by the inclusion of English Nature and other environmental bodies. Similar regional coastal defence groups have since been formed around the entire coastline of England and Wales, with the encouragement of the Ministry of Agriculture, Fisheries and Food (MAFF), the body with overall responsibility for Government policy and funding of coastal defence. Encouragement, including substantial grant aid, is being given to the preparation of shoreline management plans by groupings of authorities according to cells defined by natural coastal processes, and these plans will provide the essential context for the implementation of coastal defence schemes.

In 1989 the Solent Water Quality Conference (a Standing Conference) was formed to bring together the local authorities, who have an important role with regard to public health, leisure and recreation, and tourism, in order to raise the profile of water quality issues and increase pressure for improvements to standards and facilities. Southern Water Services, who are responsible for the collection, treatment and disposal of waste water, and the National Rivers Authority, the regulators' authority, have been key players in the Conference. A more co-ordinated approach to water quality sampling and the resolution of water quality issues has resulted.

Another important move came with the publication by Hampshire County Council of A Strategy for Hampshire's Coast in 1991. The County Council had taken a growing interest in coastal issues throughout the 1980s, having formed a Coastal Conservation Panel to oversee the acquisition and management of its growing portfolio of land holdings acquired for coastal conservation and recreation (totalling more than 1,500 hectares in 1996), and to take an over-view of the coastal defence situation. The County Council's concern for coastal issues led to the establishment of a dedicated post responsible for the preparation and implementation of the County's coastal strategy, the first of its kind in the country. A Strategy for Hampshire's Coast went beyond the confines of statutory planning, covered a wide range of management issues and argued for an integrated approach to the planning and management of coastal land and related inshore waters.

By its very nature the Strategy had limitations, because it was confined to the policies and views of a single authority: the County Council. However, it opened the debate about the growing problems of the Solent, at a time when the national debate about coastal zone management was coming into full swing.

In the early 1990s some important events occurred at national level. Whilst it is outside the scope of this paper to go into detail it is worth mentioning the most significant events, because they provide the broader context and in some respects the impetus for what was happening in the Solent.

A major landmark was the publication of the House of Common's Environment Committee report on Coast Protection and Planning. The report embodied many of the major criticisms levelled by environmental organisations and others at the system of coastal planning and management in England and Wales, and it contained some strong recommendations for change. The eagerly awaited Government response was muted. Whilst the Government acknowledged several of the shortcomings it considered that a radical overhaul was unjustified and that minor improvements to the system would suffice.

Four main initiatives resulted:

- the formation of the national Coastal Forum, serviced by the Department of the Environment (DoE);
- the publication by the DoE of Policy Guidelines for the Coast in November 1995;
- the publication of a Best Practice Guide in October 1996;
- a review of byelaw powers on the coast.

In 1992 the Government published Planning Policy Guidance Note No. 20 (PPG2O - Coastal Planning) the first comprehensive guidance concerning statutory planning on the coast. This encouraged local planning authorities to ensure that coastal issues were given proper consideration in the preparation of structure plans, local plans and unitary development plans, and even to consider the need for preparing non-statutory plans through joint working arrangements.

The preparation of non-statutory plans was being given impetus from other directions, too. In particular, English Nature had carried out a major inventory and assessment of the importance of estuaries and had concluded that the best way of ensuring their long-term conservation was through the preparation of estuary management plans. These plans were to be jointly funded, and prepared through widespread consultation. Plans of this type are being prepared in the Solent for the Western Yar and Medina estuaries, and for Portsmouth and Langstone Harbours, through partnership arrangements between relevant organisations.

One of the main recommendations in Hampshire County Council's coastal strategy was that a broadly based co-ordinating agency should be formed for the Solent. It was acknowledged that co-ordinating bodies had been formed for particular sectors, as already mentioned - for example SCOPAC (for coastal defence), and the Solent Water Quality Conference - but there was no umbrella body looking at all aspects of the Solent. Following publication of the strategy Hampshire County Council held informal meetings with some of the key players in the Solent and found that there was substantial support for the idea of an umbrella body. As a result, the Solent Forum was formed in December 1992.

The aims of the Solent Forum are to:

- facilitate more integrated planning and management for the Solent;
- assist the authorities and agencies in carrying out their functions;
- provide a 'voice for the Solent'.

The objectives of the Solent Forum are to:

- provide a broadly-based consultative forum;
- raise awareness and understanding;
- contribute to policy development, plan-making and strategy formulation;
- improve the information base;
- facilitate better communication, consultation and liaison;
- comment on major proposals for development and any other changes likely to materially affect the Solent, if the opportunity arises;

- promote the national importance of the Solent.

There are around 50 statutory and non-statutory bodies represented on the Solent Forum, including local authorities, harbour authorities and commercial, recreational and environmental organisations (virtually all of the authorities and agencies listed in Table 9.1 are represented).

In November 1994 the Solent Forum agreed to prepare strategic guidance for the Solent in two phases. Phase 1 involved the gathering and preliminary assessment of information, co-ordinated by Hampshire County Council and using appropriate organisations to provide information. The phase was completed in March 1996 with the publication of 'Towards Strategic Guidance for the Solent: Summary of Information'. For Phase 2, a Solent Project Officer was appointed to prepare the strategic guidance in close consultation with organisations on the Forum. The Project Officer post is jointly funded by twelve authorities and agencies, and its priorities and work programme are guided by a Steering Group of representatives of the funding bodies.

The consultation draft Strategic Guidance for the Solent was published in May 1997 and discussed at a conference on "The Future of the Solent" held in Southampton on 24 June. The comments received will be considered carefully and the Guidance revised as necessary.

The purposes of strategic guidance are to:

- provide a framework for addressing planning and management issues of strategic importance;
- bring together a wide range of authorities, agencies and interest groups;
- develop agreed policies to protect the economic and recreational importance of the Solent while conserving the natural resource and historic heritage;
- influence the preparation and review of statutory development plans with regard to relevant planning and management issues;
- set out an action plan for implementation;
- provide a clear statement of the implications of long-term sustainable use of the Solent, and of the means for measuring progress.

It is recognised that the guidance will be non-statutory and could be ignored. However, the process of bringing together a range of organisations, often with competing interests and conflicting objectives, to shape strategic policies is in itself an effective means of resolving planning and management issues. It is true to say that the process is as important as the plan.

Not all issues will be resolved through the preparation of strategic guidance but it will go a long way towards achieving a balanced approach which augers well for long term, sustainable development of the Solent.

References

Department of the Environment 1992. *Coastal Zone Protection and Planning.* (The Government's response to the report from the House of Commons Select Committee on the Environment).

Department of the Environment, 1992. Planning Policy Guidance No. 20 *Coastal Planning.* HMSO

Department of the Environment, 1995. Policy Guidelines for the Coast.

Hampshire County Council, 1991. *A Strategy for Hampshire's Coast.*

Hampshire County Council, 1993. *Mooring Studies and Boat Counts in the Solent: Summary and Analysis.*

Solent Forum, 1994. *Solent Forum Directory.*

Solent Forum, 1996. *Towards Strategic Guidance for the Solent: Summary of Information.*

Solent Forum, 1997. *A Future for the Solent (Conference proceedings).*

10 Re-introduction of salt marshes: preserving the coastline

R.K. BULLARD

Introduction

Over the millennia successive generations have attempted to keep the sea from invading the land, or to put it another way, to keep the coastline where it is. Some have won back the land from the sea and the Dutch must be considered the most successful in this process. Others have been less successful, which has resulted in the loss of the land to the sea, sometimes with catastrophic impacts and the loss of large amounts of a coastline.

The study of a country's coastline in many parts of the world is a study of man's ingenuity over nature. Some countries are blessed with a relatively permanent coastline as nature provided; the hard, semi-permanent, natural variety. Others have the soft, less permanent coastline that nature provided, and the local inhabitants have been battling with it since they settled in the area and started the process of surviving with the elements that they inherited.

The coastline has and continues to attract the majority of a country's population. It goes without saying that such a statement does not apply to a land locked country, but even in inland countries rivers and lakes provide similar attractions to a larger population density. Ease of access, fertile lands, abundant supplies of fresh water, are many of the factors that attract a community to live on a coastline. While these are all positive factors, a number of negative ones occur, namely, flooding and storm damage, warmer and more humid weather conditions than those usually experienced in the interior (this especially applies in the tropics), and more incidences of malaria and other diseases, many water borne, occurring due to stagnant pools of water in swamps and other flat areas. The difficulty with sanitation and the moving of liquids over shallow gradients add to the negative factors. Pollution occurs, both wind and water borne especially where the sea is used as a big dumping ground. Only now are societies having to realise that the sea cannot withstand the onslaught of dumping, nor can its inhabitants sustain the removal of species to extinction.

The concern for all those residents living on the coastline is that they can live in peace and continue with their lives without the ever present fear of the coastline collapsing or flooding. An extreme example is that more than sixty per cent of The Netherlands is below sea level. Without adequate protection the majority of the Dutch population would live a constant nightmare, what would happen in the event of the coastline collapsing? For those persons who have seen the polders, and the drainage facilities to keep the waters out, would appreciate the continuous ongoing task to keep the waters out. The Dutch Coastal Management Boards are one of the oldest of their type in the World, and have an important position in the land and water management activity.

In its attempt to seek public opinion, the Department of the Environment (1993) distributed a paper on 'Managing the Coast', where it stated that

> effective management of the coastal zone requires the integration of economic and environmental considerations.

This statement is becoming increasingly important as the changes, including global warming, have a direct bearing on the coastal zone. Boorman (1989) had earlier covered similar ground on the impact of climatic change causing rising sea level on the British coastline.

The changing coastline

The impact of the wind, the waves, and the tides, are some of the forces that are continually changing the coastline. Without man's interference and the inevitable conflict of interests, namely what coastline if any would be protected, and which should be abandoned to nature, these considerations will always enter the equation. There will always be priorities for defending the coastline, and the option 'leaving it to nature' might not be considered let alone accepted by the persons making the ultimate decisions. Nor are they likely to be accepted by those living on the coastline, and/or those with an interest in its retention.

The coastline should ideally be in a balance, that is, the forces of nature should be accommodated without damage of a significant amount occurring.

While the severe conditions, exceptional storms coupled with low barometric pressure, onshore winds, and spring tides, cannot be coped with except at considerable cost and effort, the average forces of nature should be controlled in a managed way.

It has to be accepted that not all of a country's coastline can be protected to the same extent. Less populated areas, and less important areas to the national economy will be given less protection. In some situations no protection at all. It is the opinion of the author that this should not mean that a persons property, and even their lives should be sacrificed because of this approach. With isolated properties perched on cliff faces, the owners should be compensated and re-housed at the States expense. This is of course only a general statement and every case should be treated on its merits. Life and property should be defended, at reasonable cost, but where this cannot be achieved alternative finding should be provided.

Forces at work

There are a number of forces that are acting on the coastline, and while many of them are natural many of them are often aided and abetted by human intervention. The forces listed below are impacting on a coastline that is often altered by man in an attempt to protect it from some or all of them. Sometimes the action of defending the coastline in one place can have devastating consequences, certainly at another place, if not at the very point that is being protected. The major forces are wave action, sea level changes, land level changes, coastal erosion, wind erosion, accretion, collapsing cliffs

Conflicting uses of coastline

There has been an increasing awareness of the uses to which the coastline can be put, and hence a realisation that there will be a conflict of interests. It has been recognised that there needs to be a coastal management strategy as increasing economic demands beset the coastline. Conflicts will arise between the conservationists, those with an established interest. Amongst the latter will be the residents, the industrialists, the fishing industry, the harbour facilities, and the pleasure industry, including marinas, piers and other facilities.

The leisure industry, which covers tourism is one of the fastest growing activities and provides increasing employment, and gross national products for all nations and in the Caribbean the figure may reach more than 40 per cent. Dumahie (1997) lists a number of items of good practice that she stated should

167

be adhered to when considering coastal tourism. They include:

- controlling visitor numbers within a capacity limit;
- involving the local community in decision making about development;
- formulating environmental policies covering habitat protection and water pollution;
- investing in infrastructure, transport and water management;
- exercising land use planning and development control;
- monitoring tourists and their spending, encourage good quality tourism that will be compatible with the cultural identity of the host community.

While the above were presented for coastal tourism, most apply to all users of the coastline, and those that do not only need a word or two changed to adapt them for other users.

In a recent conference entitled "The Development and Management of the Coastal and Marine Estate in the UK and Europe", the following were considered the four most important industries from a developer's point of view according to Smith (1997). They were ports and waterfronts, offshore oil and gas industry, the leisure industry, fisheries including fish farming.

Ownership of coastline

In Europe responsibility for the coastline is largely the responsibility of the central government with this authority delegated to water authorities and other bodies. In the United Kingdom, the Crown Estates own about 55 per cent of the foreshore, and almost all the seabed out to the territorial limit, and with the mineral rights of the continental shelf but excluding hydrocarbons.

The National Trust is a charitable body responsible for the purchase and subsequent management of land and improvements of major importance to the country's heritage. The NT obtains its finding from its membership fees together with the income from entrance fees to its properties. One of its greatest assets, and one which is increasing in size, is the 'coastal estate'. The NT has acquired nearly 900 kilometres of coastline in England, Wales and Northern Ireland.

Responsibility for the coastline is further complicated by the jurisdiction of various organisations that have considerable overlap. While the attention is to the coastal zone below the High Water Mark there are many organisations inland that have an impact on the coastline. Many of these are causing environmental damage, with particular impact on the salt marshes.

Rigidity versus flexibility

As stated earlier the geology of a country will determine whether there is a rigid or flexible coastline. To the coastline can be built protective measures for its preservation, and these can be of a rigid or flexible nature. The UK can be divided into four types for the purpose of coastal zone management (Smith 1997), listed in Table 10.1.

The soft option

While there must always be a place for rigidity, the quayside at a dock is an obvious example, this paper is considering the soft option where flexibility can be introduced. The salt marsh can be defined as

> an intertidal area of fine sediment stabilised by vegetation, occurring extensively along the seaward edges of low-lying coastal areas (Boorman 1995).

The process of reclaiming coastal marshes by endyking, using the natural ability of the coastal systems to build up the protected marshes by warping before finally excluding the sea has been going on for centuries (Clayton 1993). With rising sea levels this process is unlikely to be continued.

In a paper on salt marsh restoration, Casagrande (1996) looked at the likely impact on the persons living near the site and what they wished to gain from the activity. While it is recognised that the site was not directly concerned with coastal protection this was a factor that was considered. The persons were interested in restoration if it could improve their passive activities, walking, relaxing and enjoying the views. While this was a North American survey, it was interesting to note how the active pursuits like boating, canoeing and swimming scored the lowest ranking. Before change had taken place they perceived that the salt marsh was highly polluted by garbage, altered by development, and somewhat dangerous due to crime. The site had grown tall reeds, which provided a shelter for potential criminals, the restored salt marsh contained far shorter grasses. They wanted the restored salt marsh for its natural beauty and the wildlife that it contained more than they valued it as a resource for exploitation.

Table 10.1 The main characteristics and coastal hazards of coastal types

Coastal type	Main characteristics	Natural hazards
Northern	Hard rock coasts with deep fjords, immense variety of coasts, rock types and structures. Geologically stable or mainly uplifting	Strongly dependent on morphology and geology of coast. No/little risk from volcanoes, earth-quakes, and sea level rise
Western	Rocky shorelines with many inlets (rias) and frequent extensive beach/dune complexes	Strongly dependent on morphology and geology of coast. No risk from volcanic activity
Central	Low sedimentary coast with mudflats, marshes, beaches, barrier islands, wadden seas and estuaries. Geologically subsiding	High risk of flooding, storm surges, and sea-level rise, dependent on coastal morphology, and physiography. No risk from volcanic activity
Southern	Varied landforms. Strong tectonic influence	Risk from volcanic and earthquake activity. No risk from storm surges

Source: Smith (1997)

East Anglia - a case study

The East Anglian coastline has a well-known history of erosion especially in such examples as the town of Dunwich on the Suffolk coastline (Clayton 1993). It suffers from many of the problems mentioned above, the sinking land, rising sea level, unstable cliffs where they exist, sand banks, onshore wind, and cross currents. The examples presented here follow the example of an attempt by to protect the coastline with offshore reefs, previously written about for the 22nd International Symposium in Olstyn, Poland (Bullard,1995). This example which is being constructed by the National River Authority (NRA) has continued to be monitored by the author, and shows that offshore reefs may not be the solution they were intended.

In Essex, the coastline is protected by mudflats, saltmarsh, sea walls, and coastal grassland (King & Bridge, 1994). It contains 17 SSSIs (Site of Special Scientific Interest), 5 NNRs (National Nature Reserve), 4 LNRs (Local Nature Reserve), 3 SPAs (Special Protection Area) and Ramsar (Ramsar Convention for designated wetlands) sites. The Blackwater Estuary has a Management Plan, which has as its main coastal issues recreation conflicts and access, saltmarsh erosion, international importance for wildlife, sea level rise, and increased storm surge frequency.

An attempt by English Nature to reintroduce salt marshes at Tollesbury on the Essex coastline is a further attempt to protect the coastline on an experimental basis. The example shows that an attempt to re-introduce salt marshes can only be undertaken with the support and financing of the owners of the coastal zone. In this example, the land is owned by Wilkins the jam makers. The strategy was to withdraw land from farming in this example the growing of strawberries, and to receive the appropriate 'Set Aside' compensation.

The site has retained many of its original features, the extensive hedge and row of trees. There has been a further bund constructed at the foot of the farm land, this is considered an unnatural feature. The original bund was retained but was breached with a gap of some 20 metres. Of interest is that this calculated gap has succeeded because little subsequent erosion or deposition has occurred since the breaching occurred. The twice daily tide is well accommodated.

Of concern if the lack or limited vegetation that has become established on the reintroduced salt marsh. With surrounding salt marshes it was hoped that there would be a natural expansion, this has not occurred. Planting would be an alternative, the site is already unnatural, so more influence by man would not be a problem. Plants will tend to grow at different levels (Penning-Rowsell *et al* 1992) and with a breech situation being imposed on the site certain species may not obtain a foothold. The site will need regular monitoring and the use of ground measurements together with aerial photography and satellite imagery should be considered (Vaughan, 1994). A paper by Pethick (1996) shows clearly that the monitoring of salt marshes requires annual observations over a ten year period to enable progressive change to be established.

The NRA considers that by using salt marshes the crest defence walls (Table 10.2) are required to provide the prescribed standard of sea defence.

With the relative costs for the walls at a ratio of 2:4:25 it can be assumed that the introduction of salt marshes, where their introduction is possible, may result in major cost savings, the difference between building and planting. The loss of land factor must be taken into account. With the savings of constructing a wall this will provide an opportunity to value the land occupied by the salt marsh, this land may therefore become more valuable than when it was used agriculture and when it has been taken out of production. This factor may be further improved if 'Set Aside' status can be allocated.

Table 10.2 Recommended wall heights for saltmarsh protection

Saltmarsh length (metres)	Wall height (metres)
80	3
30	5
0	12

Source: NRA

Conclusions

The following issues are of relevance with regard to the re-introduction of salt marshes:

- that the cost should be less than the value of the land in its present use, considering the cost of defence walls as an alternative;
- that the rising sea level may prevent the salt marsh from becoming established;
- that regular monitoring by a selection of methods should be undertaken at least yearly over a minimum of 10 years;
- that pilot sites should be more typical than the present example;
- if natural reseeding does not occur assistance should be given;
- more attention given to environmental impact of coastal defences.

References

Boorman L.A., 1989. *Climatic Change, Rising Sea Level and the British Coast*, Institute of Terrestrial Ecology, Natural Environment Research Council, HMSO., London.

Boorman L.A., & Hazelden J., 1995. Saltmarsh Creation and Management for Coastal Defence. In: *Directions in European Coastal Management*, eds. Healy M.G., & Doody J.P., Samara Publishing, Cardigan, U.K. 175-183.

Bullard R.K., 1995. Environmental Impact of Constructing Sea Defences in a Coastal Zone, European Faculty of Land Use and Development, 22nd International Symposium on *Ecodevelopment of Rural Areas*, University of Agriculture and Technology, Olsztyn, Poland, 8th - 10th June 1995.

Casagrande D.G., 1996. A value based policy approach: the case of an urban salt marsh restoration, *Coastal Management*, Volume 24, Number 4, October-December 1996, 327-33 7.

Clayton K.M., 1993. *Coastal Processes and Coastal Management*, Countryside Commission, Cheltenham, U.K.

DoE., 1993. *Managing the Coast - A Review of Coastal Management Plans in England and Wales and the Powers Supporting Them*, Department of the Environment and Welsh Office, London.

Duniashie D., 1997. The benefits and pitfalls of coastal tourism, *Chartered Surveyor Monthly*, 30-31.

Healy M.G., & Doody J.P., 1995. *Directions in European Coastal Management*, Samara Publishing, Cardigan, U.K.

Jones P.S., *et al* 1996. *Studies in European Coastal Management*, Samara Publishing Limited, Cardigan, U.K.

King G., & Bridge L., 1994. *Directory of Coastal Planning and Management Initiatives in England*, on behalf of The National Coasts and Estuaries Advisory Group, Countryside Commission, U.K.

Penning-Rowsell E.C., *et al* 1992. *The Economics of Coastal Management - A Manual of Benefit Assessment Techniques*, Belhaven Press, London.

Pethick 1., 1996. The sustainable use of coasts: monitoring, modelling and management. In: *Studies in European Coastal Management*, eds Jones P.S., *et al.*, Samara Publishing Limited, Cardigan, U.K. 83-92.

Smith H., 1997. Important Issues in UK Coastal Management, *Chartered Surveyor Monthly*, 48-49.

Vaughan R., 1994. *Recent Advances in Imaging Coastal Zones*, EARSeL Workshop on Remote Sensing and GIS for Coastal Zone Management, Rijkswaterstaat Survey Department, 24-26 October 1994, 339-48.

11 Capacity in coastal zone management at the University of Portsmouth

J. TAUSSIK

Introduction

The University of Portsmouth is particularly well placed to develop expertise in coastal zone management. It is the only British university located on an island, though it has to be said that the natural characteristics of Portsea Island, which accounts for most of the administrative area of the City of Portsmouth, were exaggerated when the northern defences of the island were constructed and a clear defendable channel cut. As the home of the British Navy for several centuries, the city had a range of industries associated with the sea. Its long tradition of maritime building and engineering, combined with the presence of a number of historically unique ships, underlies Portsmouth's naval heritage.

Portsmouth lies in the centre of the South coast of England. Portsea Island faces the Solent, an estuarine complex running from Hurst Spit at its Western end to East Head at the Eastern end, extending Southwards to the Isle of Wight (see Figure 11.1). Formed when the drowned, east flowing Solent River breached the chalk ridge, which linked the Isle of Wight to the mainland after the last Ice Age some 100,000 years ago, the Solent contains a rich variety of coastal forms. It includes three major harbours (from West to East: Portsmouth Harbour; Langstone Harbour; and Chichester Harbour) stretching some 10 miles inland. Parts of the coast remain unspoilt and offer a range of landscapes protected by national and regional designations. The Solent is located at a transitional point between different biogeographic realms resulting in the existence of a rich variety of species, often at the margins of their natural range and, therefore, highly fragile. It boasts significant levels of certain habitats, e.g. brackish lagoons and salt marshes, and high proportions of certain species of migrating birds, e.g. Brent Goose. The numbers of international demonstrates this biological richness and diversity, national and regional designations protecting wildlife and their habitats. The area has been inhabited since earliest times. Its rich archaeological and historical heritage, combined with its natural attractions, makes it a popular tourist destination.

The sheltered nature of the estuary, combined with its particular tidal regime, has given it natural advantages for port development. Southampton and Portsmouth are international ports, significant for passenger travel and trade. These same advantages have made the area particularly attractive for a wide range of recreational activity and the area is particularly renowned for its sailing facilities and tradition. The draft document, *Strategic Guidance for the Solent* (1997), provides additional information about the area.

The local coast, therefore, offers considerable scope for coastal and marine research and study, a benefit maximised by the University. Courses as disparate as biology, geography, geology, leisure resource management and media studies use the coast as an outdoor workshop or studio. A wide range of research is undertaken, from marine biology, with its laboratory on Hayling Island, to scientific analysis of pollutants to furthering understanding of coastal processes along the Solent. From this established local base, the University has established an international reputation in a number of areas of research and its expertise is sought around the world.

Coastal management in the Solent

While the economy is diverse, the area is recognised as providing a centre of excellence for marine industries, science and technology. Currently, there is a resident population of around 1 million in the immediate coastal area of the Solent. Such levels of human activity in the coastal zone generate enormous pressures and problems on the coast: physical processes of sediment movement and coastal erosion patterns have been modified by coastal defence and dredging activities; habitats have been lost through land reclamation; water quality has been threatened by waste and pollution discharge; and wildlife has been disturbed. Always there is the treat of catastrophic oil pollution (Solent Forum, 1997).

It has been suggested that the area is unique in the level of competition that exists for coastal resources (Solent Forum, 1997). Such competition inevitably brings conflict and it has long been recognised that an integrated approach to the management of the Solent is necessary. Local authorities, particularly, in the area have responded to this need by establishing frameworks for the development of management policy at different levels and they provide leading examples for other parts of the UK (Taussik, 1996). The paper, *Partnership in Planning and Management of the Solent* has described the nature of these local approaches to reducing that conflict. The draft Strategic Guidance (Solent Forum, 1997) for the area refers to 'balance' in the Solent and suggests that the long term planning and management for the area should seek to sustain this.

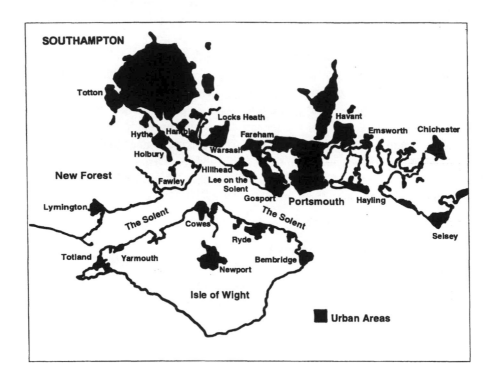

Figure 11.1 The Solent region

The Solent, therefore, provides a further laboratory for the University in terms of the management and regulation of the local coast and its resources. This work includes analysis of the institutional systems affecting the operation of policy related to coastal resources, development of methodologies for valuing a range of coastal resources and comparative assessment of public participation in coastal zone management.

It is clear, therefore, that the University of Portsmouth is fortunate to be located in such an enormously rich and varied coastal environment, an opportunity which it has maximised in its research and which is reflected in the nature of the teaching offered. In fact, a informal trawl of University research a few years ago revealed some 80 individuals or small groups working on coastally related projects. This covered a wide spectrum of expertise and means that groups can be drawn together to work on specific projects of very different character. With this complementarity, it is not surprising that the University has identified maritime studies (in its widest sense) as offering opportunity for further growth. This will provide distinctiveness in its courses and expertise in the broad spectrum of higher education.

Coastal research

In the context of the theme of this conference, Coastal Zone Management, this paper will concentrate on the three of the University's research groups most closely connected to management aspects of the coast. It will then outline their related postgraduate courses.

The three research groups are the:

- Centre for Economics and Management of Aquatic Resources (CEMARE);
- River and Coastal Environments Research (RACER);
- Centre for Coastal Zone Management (CCZM).

The Centre for Economics and Management of Aquatic Resources (CEMARE)

CEMARE is based in the Economics Department of the Portsmouth Business School. It was established in the early 1960s to promote interdisciplinary research into marine resources and has developed into a substantial centre for research and consultancy in fisheries economics, aquaculture and marine and freshwater fisheries management. Its interests extend over both the commercial and recreational sectors. It has developed an international reputation through several continents although it still undertakes valued work in the local area. Major areas of work include; European Union fisheries, aquaculture and artificial reefs, fishing enterprise management and decision making, bio-economic modelling, and risk analysis.

In addition to its research and consultancy interests, CEMARE is also involved in training and education. It has organised regional, national and international conferences concerned with fisheries economics as well as running short courses on a range of topics. It has produced a substantial number of its own CEMARE papers.

River and Coastal Environments Research (RACER)

RACER comprises a team of coastal and fluvial geomorphologists operating within the Department of Geography at the University. Established in 1988, the group has a sound record of research and consultancy covering both natural processes and landforms as well as developing the management applications of such knowledge. Projects range from detailed investigations of localised problems to large-scale strategic studies. The team has done a considerable amount of work for SCOPAC, The Standing Conference on Problems Associated with the Coast. This group brings together those with interests particularly in coastal defence for the Solent area. This includes, for example; sediment transport and budgets, littoral cell identification, impacts of sea level rise; and provision of tidal and sea level information.

Fundamental to their work has been the translation of geomorphological principles and the understanding of processes to provide clear guidance for coastal managers. Of special importance is an appreciation of the extent of the modification of natural processes by particular management practices and policies in terms of their wider and longer term impacts.

The Centre for Coastal Zone Management

The newest of the three, but with the widest remit, the Centre for Coastal Zone Management, was established in 1992 as an inter-faculty, inter-disciplinary research group. It involves staff from the Departments of Land and Construction Management, Geography and Economics and its expertise includes a wide range of social scientists: economists; planners; land managers; geographers; political economists; valuers; and environmentalists. The establishment of the Centre reflected recognition that coastal zone management was a necessary response to the increasing pressures on coastal resources, locally, nationally and world-wide. Its purpose is to undertake research into techniques and methodologies used by resource managers and to explore, particularly, their application to coastal resources. Current areas of interest

178

include: planning in the coastal zone; the valuation of coastal resources; institutional aspects of resource management; and public participation in coastal zone management.

To further its research in coastal zone management, CCZM has undertaken or contributed to a number of consultancy projects, including several estuary management plans. Other outputs include its own Working Papers in Coastal Zone Management series, a substantial number of papers published in academic journals or presented at international conferences and the organisation of regional, national and international conferences. The Centre was particularly pleased to organise LITTORAL 96, the third international conference of EUROCOAST, at Portsmouth in September 1996.

It is clear from the range of interests involved in CCZM that the group is interdisciplinary. This reflects general thinking that coastal zone management must integrate across all disciplines, sectors and fields of management operating in the coastal zone (OECD, 1993). It also reflects the fact that techniques and methodologies of value in coastal management will come from a range of different disciplines, for example, from: political science; valuation and economics; town and country planning; and sociology. This inter-disciplinarity is considered to be a strength of the group. Its particular combination of expertise is perceived to fill some of the gaps in current approaches to coastal zone management. However, current approaches to the distribution of research funding stresses the role of traditional disciplines, which can make interdisciplinary work appear peripheral. Greater recognition of the importance of interdisciplinary research is required to ensure that appropriate resources are made available which will allow new solutions to be found for emerging problems, both in the coastal zone and in other areas of resource management. This is particularly important, given what West (1997) says of coastal management:

> This profession (coastal management) is driven by problems which more often than not have been recognised outside of academic institutions, the successful solution of which has required input from a number of conventional disciplines. Problem solving in coastal management increasingly will require the expertise of members of the physical, biological and social sciences for their successful development and implementation.

Synergy between the coastal research groups

These three groups, CEMARE, RACER and CCZM, encompass the coastal researchers most concerned with coastal management issues. Although each group has a particular emphasis and has established national and international reputations, there is considerable synergy between the groups, which can be exploited to respond to particular research or consultancy contexts. For example, economists and geomorphologists can work co-operatively on problems concerned with evaluating sea defence proposals (including 'do nothing' and managed retreat) or planners and geomorphologists can work together on shoreline management strategies. The University has additional, highly valuable expertise in Geographic Information Systems (GIS) and remote sensing which complement and support a whole range of other areas of work including coastal management.

Postgraduate courses

The three research units have been very successful in developing national and national reputations in their field and the opportunities have been taken to make available that expertise to students (see Figure 11.2). For example, the staff of CEMARE are associated with a suite of named Masters awards in Fisheries Economics. Closely linked, this provides for awards in; fisheries management, fisheries economics, recreational fisheries management, and fishing enterprise management.

Members of RACER and CCZM teach on a range of Masters level programmes in the Faculty of the Environment including Environmental Resource Management, Land Information Management and Mapping and Ecotourism but the main vehicle for developing coastal expertise is through the Masters programme in Coastal and Marine Resource Management.

MSc Coastal and Marine Resource Management

The Masters programme in Coastal and Marine Resource Management has been running for three years and reflects the intention to broaden the output of the research of CCZM. While the course is organised from the Department of Land and Construction Management (LCM), input to the course comes from the Departments of Geography and Economics as well as LCM, reflecting contributions from each of the research units: CCZM, RACER and CEMARE. Such an interdisciplinary, cross faculty structure is typical of coastal management programmes in the United States (West, 1997), where coastal zone management has a much longer and more secure history and where, therefore, there is a much longer tradition of university education in coastal management.

Finkl (1997) states that

> the educational process (in coastal management) must be integrative and responsive to long term societal needs.

The course has been developed to increase understanding of resource management in the coastal and marine environments. While certain approaches and techniques are widely applicable, the context in which either coastal or marine resource management operates is substantially different from the context affecting land-based resource management. Considerable attention is, therefore, paid to the physical and social context of dealing with the management of these resources. The intention is that, by integrating the economic/evaluative aspects with the principles of resource management and policy formulation in the context of coastal and marine environments, students will develop a broad and deep understanding of the exploitation and management of coastal and marine resources. Students are introduced to a wide range of management techniques that are transferable to a range of contexts. The intention is that they should be given tools, which will allow them to derive solutions to unfamiliar problems. The importance of the integration of valuation, management, law and policy elements in the course has been recognised by the accreditation of the course by The Royal Institution of Chartered Surveyors. The course, therefore, provides the academic qualifications necessary for entry to that profession. It has also been recognised by the Department for International Development (DEFID) (replacing the Overseas Development Administration) as providing preparation for working in coastal zone management anywhere in the world.

The course encourages self-development and independent learning and the programme has been devised to operate as flexibly as possible for students from a range of backgrounds. For example, the traditional study route can be studied by full or part time attendance. An alternative independent study route is available to students who have a well-established history in coastal zone management. This route switches the balance of study to emphasise student centred learning and research and has been used for both distance learning and with attendance at Portsmouth.

The course has proved very popular with students. They are concerned about the escalating pressures, and resulting problems, on coastal resources. While they appreciate the role of science in underpinning decision making, they recognise that the approach to the long-term sustainability of coastal resources must come through management regimes. They value the inter-disciplinarity of the course and their introduction to a wide range of concepts and techniques.

The course benefits from the wide international network of contacts of the members of CCZM and some of the students have taken advantage of this during or after their study. It is hoped that these links will be extended, providing opportunities for exchange between students or staff of different countries and demonstrating the significance of coastal zone management at an international level.

Conclusions

Located in a city with a long maritime history and in a region with a coast that is physically and socially complex, the University of Portsmouth is particularly well placed to develop expertise in coastal and marine studies. It has exploited this opportunity on a wide range of fronts, including in the management of coastal resources, through research and teaching.

The interdisciplinary nature of coastal zone management is reflected in the membership and research of CCZM, the broadest ranging of the three research groups involved with the management of coastal resources. Following its success, the Masters programme in Coastal and Marine Resource Management has been developed as an inter-faculty, inter-disciplinary initiative in response to perceived needs in the UK and internationally for resource managers with particular knowledge of the coast. The success of the course to date, and its recognition by external bodies, suggests it has a sound future. In its further development, it is hoped that the opportunities for international linkage will be maximised.

Most authors, whether writing about the activity of coastal zone management or about education in coastal zone management, stress the importance of integration and the need, therefore, for interdisciplinarity. Students wishing to pursue careers in coastal zone management similarly recognise the importance of this feature. The importance of interdisciplinary research, not only in coastal management, must also be recognised by the research community to ensure that such areas are appropriately resourced in the future. West (1997) states:

> If these kinds of research areas are not permitted to flourish, it will eventually stifle not only the academic community but result in a reduction of viable solutions to problems in the field.

References

Finkl, C. W., 1997. The coastal zone as a new battlespace from the purview of an academic graduate program in environmental science. In: Miller, M.C., & J. Cogan. *Coastal Zone 97. Abstracts of Presentations.* Boston, USA: Coastal Zone 97.

Organisation for Economic Co-operation and Development, 1993. *Coastal Zone Management: Integrated Policies.* Paris: OECD.

Solent Forum, 1997. *Strategic Guidance for the Solent. Consultation Draft May 1997.* Winchester: Hampshire County Council.

Taussik, J., 1996. Development plans and the coastal zone. *Town Planning Review.* **67**(4), 397-419.

West, N., 1997. The education of coastal managers: The role of the universities. In: Miller, M.C. and J. Cogan. *Coastal Zone 97. Abstracts of Presentations.* Boston, USA: Coastal Zone 97.

12 Lytham-St. Annes: the decline of a British seaside resort

D.C. DOUGHTY

Introduction

The popularity of British seaside and health resorts and Spa towns in the eighteenth and nineteenth centuries led to a development of holidays for all. The coast of western Lancashire became a playground for the rich and for the workers of the inland textile towns. Lytham had been a fishing village and resort for many years before the development of the new health and luxury resort of St. Annes on the Sea further along the Ribble estuary. The new town flourished and gave the coast an element of class distinct from the rowdy nature of nearby Blackpool. The advent of package tours and cheap holidays to Spain and other sunshine destinations in Europe in the sixties soon led to a decline in the resort, exacerbated by the wishes of the post war retirement classes to settle in the town and create a privileged ghetto of old age pensioners. Further decline set in with recent recession and a glut of unsellable properties along the coast caused by the impoverisation of the middle classes after the failure of old established Lancashire capitalism. The sorry state of a resort whose buildings and infrastructure have been left to waste other than the add-ons of Supermarkets and cheap attractions in nearby Blackpool has seen the decline and fall of the once elegant health and seaside resort.

Geography of the Fylde

The geographical backbone of the British Isles, particularly the north of England, is the Pennine range of hills. The range divides the country into west of the range and east of the range and industrial developments during the nineteenth century in particular accented the two cultures of Lancashire and Yorkshire. To the east of the divide were the 'wool towns' of Bradford, Leeds and Halifax. To the west of the hills were the 'cotton towns' of Manchester, Oldham and Bury. Both Yorkshire and Lancashire had their ports at Hull and Liverpool respectively and these basic divisions, despite more recent boundary divisions still held true at the end of the twentieth century.

Not only did industry thus settle into its different ways on either side of the ridges, but leisure resorts followed their example. The Yorkshire holiday-makers tended toward Scarborough and Bridlington, the Lancastrians towards Southport and Blackpool. However, the Lancashire resorts were divided again by the geographical features of the two main rivers in the region - the Mersey and the Ribble. Thus, South Lancashire and the city of Liverpool (or Merseyside as the area is now partly called) spawned the elegant resort of Southport with its monumental buildings along the elegant shopping boulevard of Lord Street, whilst northern Lancashire and the cotton towns around Manchester gravitated towards the more tacky aspects of Blackpool.

This is not exactly quite as simple as outlined and it would be unfair to suggest that the Liverpool middle-classes were responsible for the good taste and sanity of Southport whereas the Manchester workers were entirely responsible for Blackpool. Also, travel was not so restricted and many factory owners in Manchester had their families installed in Southport rather than Blackpool and the Fylde. Nevertheless, the river Ribble certainly provided a barrier to discourage Liverpudlians from travelling the extra distance from Southport to Blackpool, which involved and still does involve a considerable detour along both banks of the river and via the industrial centre of Preston. Bridges have been proposed but the estuary suffers from the problems of sinking sands and a recent attempt at linking the Lytham side with Southport failed because of costs and lack of interest. Perhaps the most positive side of this failure to bridge the small distance across the estuary banks is that Southport and Blackpool have been able to retain their separate characters to a certain degree.

Development of Lytham and founding of St. Annes

If Southport represents what, at least, was a middle class dream seaside settlement, then the Fylde coastline also made attempts at creating settlements and resorts different from the basic working class fun of Blackpool.

Blackpool always had an advantage over Southport in attracting the Lancashire cotton workers. It had a direct rail connection to Manchester and inner Lancashire well before Southport (as early as 1830) and indeed by the 1850s the day tripper traffic from the Lancashire and Yorkshire railways had reached levels of up to 12,000 visitors in August weekends. Southport was not able to compete in this market (nor was Morecambe further up the coast) until the late 1850s (Walton, 1992).

Of the four resorts on the Fylde coast however, the oldest is Lytham and this fishing settlement predates any tourist resort in western Lancashire. The port of Lytham appears in the Domesday Book of 1086 spelt as Lidun. William Thornber in his history of Blackpool suggests the name comes from the Anglo Saxon word lethe meaning barn or, more probably a combination of lade and tun meaning a settlement on an estuary. The Fylde at this time was in Norman hands under Roger of Poitou but by 1200 had passed to the ownership of the monks of Durham priory who established shrines to Sts. Cuthbert and Mary. In 1606, the lands passed to the Clifton family, remembered now in Lytham's best hotel's name - the Clifton Arms. (Ramsbottom, 1995).

In 1813, visitors to Lytham however, were arriving in 'shoals' to enjoy the salt waters even though the town is properly only still on the banks of the Ribble river estuary rather than the open sea. By 1846, when the Preston to Wyre railway opened a branch line from Kirkham, the village had already developed into a seaside resort, a quieter and more sedate alternative to the hustle and bustle of Blackpool a few miles up the coast. The development of Lytham (and indeed of Blackpool) was thus heightened by the new railway which made both places more accessible for the inhabitants of Preston and Lancashire, than nearby Southport and which was to lead to the rapid expansion of building in Blackpool and to some extent in Lytham, although the control of the Lytham manor by the Clifton family, did much to stop the haphazard sort of development that was to plague the larger resort (Granville, 1841).

Unlike its twin resort of Lytham, St. Annes is a relatively modern settlement, which experienced a remarkably rapid growth. The first building to be constructed on the Clifton family lands north of Lytham was no more than a wooden hut built in 1875. By 1916, the town of St. Annes had grown from those humble beginnings to a seaside resort with a population of 11,033. The resort was built to a specific plan and in contrast to Blackpool and Lytham:

> St. Annes was intended for mental and bodily recreation . . . never should any person who proposed to bring into the town that which would debase its moral tone or render it a place where a family could not come and leave their children to enjoy themselves in all parts of it.

The quote is from the speech of a certain Mr. Maxwell, one of the town's dignitaries and the echoes are those of Victorian philanthropy and the Protestant ethic; this was the time of Methodists and tea-totallers who were seeking the improvement of their neighbours whether requested or not and St. Annes was seen as a model health resort for the new middle classes (Heywood, 1916).

The original plans for this genteel seaside paradise originated in the St. Annes Land and Building Company which was formed in 1874 by Elijah Hargreaves, W. J. Porritt and five other directors who leased the land from J. T. Clifton, the Lytham landowner, for 1,100 years from March 31st 1875. Development was rapid and by 1885, the resort had an esplanade with two bandstands and gardens as well as a 400 yard long pier built at a cost of seventeen thousand pounds. Sailing boats plied their way from the resort to Blackpool, Lytham, Southport and up the Ribble estuary to Preston (still at that time a working port). By 1898 a golf links was added, one of the few surviving attractions of St. Annes and now one of the major international courses (Ibid. pp. 8-12). By 1899, the pier had expanded to include a 'Moorish Pavilion' and by 1910, a 'Floral Hall' was added where up to 1,000 people were able to listen to 'the refined music of Miss Kate Earl and her orchestra'. No less significantly it was also the venue for tours of the Manchester Hallé Orchestra (Rothwell, 1978). In addition, St. Annes received two new luxury hotels, the Grand in 1897 and the splendid Majestic, with its statue of Hygeia, in 1910 (demolished for building of a block of 'luxury' flats in 1975) (Singleton, 1996).

The flourishing health resort

The growth in seaside resorts in the nineteenth century was one of the reasons for the decline in inland Spa resorts and although the development of the railways gave a boost to the inland resorts, it also brought the seaside within easy reach of the industrial heartlands. The new concept of holidays for all and the 'Wakes' weeks meant that the industrial proletariat were able to enjoy the seaside and its health giving air and salt water bathing as well as the rich and the bourgeois classes. Thus, the coastal resorts were experiencing nothing less than a boom (Walton, 1992).

St. Annes was seen from its conception as a health resort, not just a centre for entertainment like Blackpool or Southport. This 'middle class' feel to the resort saw rateable values double over the five year period from 1891 to 1896. The town had, by then, 'gained a splendid reputation as a health resort' and was particularly recommended for treatment of rheumatism, asthma, bronchitis, consumption and other chest complaints. The town did have a death rate of only 7.8 per 1,000 as opposed to a national rate of 14.5 per 1,000 (Heywood, 1916).

Health biased hotels and hydros sprang up at the beginning of the century, ready to take the Spa treatments to the coast and praise the benefits of salt water therapy (or thalassotherapy), both in Britain and on the continent (Corbin, 1988). Apart from the traditional Grand and magnificent Majestic (originally known as the Imperial Hydro) hotels, specifically health oriented establishments opened. By 1912, the Imperial Hydro Hotel could boast Russian, Turkish, Sea water and Slipper baths as well as in-house electrical treatments by a qualified physician. Massages and late dinners were available at the Southdown Hydropathic establishment. (Bowman, 1912). Smaller establishments sprang up everywhere too, such as the Dunes House Hydro, later to be converted, as were many others, into a retirement home. For the less health conscious visitor, small hotels and guest houses appeared along the sea front and the towns main arterial road, Clifton Drive, many of which still remain to cater for the remnants of today's tourist trade - the St. Ives, the Dalmeny and the Fernlea (Singleton, 1996).

In the heady days before the outbreak of the Great War, St Annes was described as:

> The lungs of Lancashire . . . have no healthier or more capacious breathing 'cell' than St. Annes on the Sea (Bowman, 1912).

Even moments before the second war, St. Annes was still functioning as a middle class health resort. The official comments of the mayor of the time, A. J. Edge praised his resort as follows:

> Should you desire the quieter holiday, you will find that we have splendid roads, smooth and lengthy promenades, numerous and lovely public and

private gardens, municipal concerts and a complete service of remedial baths and treatments (Lytham St. Annes Corporation, 1938).

St. Annes remained the health resort par excellence of the Fylde coast and would even try to remain so after the 1939-45 hostilities. Holidays and health were still needed for all and the resurgence of Victorian values after the war meant that St. Annes was ready to take its place as the seaside Spa yet again, but post war boom and changes in attitude were to have a disastrous effect on the health resort of the Fylde.

St. Annes as a retirement settlement

Even as early as the beginning of the century, St. Annes was seen as not only a health resort for the holidaymakers from inland Lancashire; it was also viewed as a healthy resort for retirement homes. Convalescence and retirement homes soon became part of the new town's landscape. Amongst the new homes was the Blackburn and District Home built right amongst the sand dunes in 1914. Today, many of the previous guest houses and small hotels have been converted into nursing homes and retirement homes, giving unqualified personnel the chance of making profits from the old age pensioners left to fend for themselves in an economically increasingly hostile environment.

Initially, St. Annes was seen by the middle classes of central and eastern Lancashire as an ideal retirement home and significantly, the political constituency remains one of the most strongly Tory areas of the country. It is also one of the most stretched areas for health treatments and social welfare because of the imbalance of the population, with most residents being over sixty if not more.

It was the post war boom in manufacturing that persuaded the newly retired to buy properties in St. Annes. The fact of its continuing 'middle-classness' attracted those from Manchester and other inland manufacturing centres to come to the healthy seaside to settle down. St. Annes responded well with its gardens, repertory theatre, hotels and restaurants and its exclusive shops in the Square. In effect, too many elderly people flooded into the town, keeping away from the hectic atmosphere of Blackpool and creating an enclave for the middle class privileged. In the 1950s and 60s house prices rose and the settlement enjoyed prosperity and security but retirement homes eventually brought their own problems to the resort. Available monies for pensioners soon began to dwindle and after the triumph of Thatcherism in the land, the very supporters of the Tory party began to feel the pinch. Manufacturing inland declined and was almost wiped out for a while. The resurgence of Manchester as a major centre of commerce was too far away and anyway would bypass the retired manufacturing classes. St. Annes was filled with the retired classes but the residents now represented the last members of that class and as they began to die off, properties saw long term non-occupancy or unsuitable types of occupancy.

Final decline?

Lytham St. Annes as the two towns became known, including also the addition of the adjoining area of Ansdell and Fairhaven now makes up most of the estuary and sea coast of the Fylde between Blackpool and Preston. In effect, the areas all run into each other as a large conurbation and continue north towards the old fishing town of Fleetwood. Supported at its base by the industry of Preston and its reliance today on British Aerospace, much of the area has declined into the post-industrial landscape that can be seen in much of the North West of the country. The glitz and glamour, tawdriness and tatt of Blackpool remains as ever and will always attract sections of the holiday making public, be it the Scots from Glasgow on the trail of cheap beer and fish and chips or the new Gay traveller on day trips from Manchester or longer stays in the homosexual seaside capital of the North or even the political party conference or Masonic 'do'. Blackpool exists almost in a world of its own (Cohn, 1997).

Lytham St. Annes with its air of faded gentility is like many coastal resorts in Britain, the victim partly of its own success, partly of the encroachment of Europe upon the once fortressed shores of Britain.

In the 1960's when the post war boom had taken hold, when a spirit of adventure and rebellion had at last found its way into the staid minds of British youth, the entrepreneurial skills of Freddie Laker and his like introduced cheap flights and above all, the package holiday abroad. The dull skies of north west Lancashire suddenly had to compete with sunny Majorca and the Costas of Spain. New money in the economy at the time sent thousands off to foreign shores and the experience claimed them for future holidays and future years. The decline of the British seaside began to come about through vagaries of the climate and the success of post war British industry.

Resorts like St. Annes soon became the downmarket alternative to the 'jet set' continental holiday. Specifically, decline led to under-investment in infrastructure and neglect of the very resources that had made St. Annes the successful resort it was. The national health reforms of 1948 put paid to many alternative therapies; British Spas like Harrogate and Buxton received official University style reports that their functions were unnecessary and overexpensive compared to new cheap medicinal drugs, following the success of penicillin and analgesics, that had been produced for the war effort. The knock-on effect soon killed off the ideas of the St. Annes hydros and health hotels. Soon after, the effects of long term pollution came into the public knowledge and the Fylde resorts lying at a river estuary as they do soon became known as some of the most polluted and dangerous beaches in the country (Wickers, 1997).

A series of somewhat perverse disasters in the town added to the decline and the lack of any municipal support or interest in restoration became a sorry fact. Worst of all of these perhaps, was the destruction by fire, in July 1974, of the magnificent pier and its Moorish Pavillion despite the attempts of forty firemen to fight the flames. In 1982 the Floral Hall was similarly destroyed by fire. The response of the council was merely to demolish the pier's seaward end, never to be rebuilt (Singleton, 1996).

Retirement homes and a policy of non-recognition of potential bust as well as boom in the housing market led to an overproduction of homes for the elderly and retirement flats. The splendid Majestic hotel, flagship of the bourgeois holiday was demolished to make way for the usual theme pub and five storey block of flats. The tram line of 1903 linking Squires gate (the Southern entrance to Blackpool) with St. Annes was discontinued and now all that remains is the ticket office which serves as a somewhat miserable tourist office on the corner of the St. Anne's square. Even the railway, once the busy link for tourists and day trippers now retains merely an unmanned hut on a shuttle line that operates only from Blackpool to Preston, a single track infrequent shuttle service (Singleton, 1992). The Roman Baths built on the seafront and opened in 1916 saw a sharp decline and lack of any investment in the 1970s and were finally closed in 1988 despite a petition of 2,000 names (Ibid. p. 110). The bandstands, once part of popular free entertainment disappeared, the seafront gardens were left to fall into disrepair and the sand dunes, once home to rare species and play grounds for children and adults alike were razed because of complaints from rich retirees on the seafront that the sand was blowing into their living rooms and causing discomfort to the Tory voting majority.

In effect, little was done after the onset of the first signs of decline to revitalise the town. The powers that be seem to have taken the fall from grace of the once prosperous health resort as a matter of course and today's view of the Square, Clifton Drive, the Pier, Seafront Gardens or whatever is a sad echo of times truly past.

All that really remains of Lytham St. Annes today is the Royal Lytham St. Annes Golf Club, a small reminder for a week or two each year of the lost glamour of the Fylde's middle class health resort. Unfortunately seaside decline is almost part of the success of neighbouring Blackpool but the crumbling gables and empty shops of St. Annes are a sad reminder of the days that once, however briefly, made this one of the north west's foremost resorts.

References

Bainridge, B., 1997. Hometown-Formby. *The Times Magazine*. Aug 2nd, 46.

Bowman, J., 1912. *St. Annes on the Sea*. London Health Resorts Association.

Coe, J., & Murphy, A., 1997. The A-Z of the seaside, *The Observer - Life*. July 6th, 8 – 57.

Cohn, N., 1997. Western Front, *The Sunday Times - Travel*, 24th Aug, 6 – 7.

Corbin, A., 1988. *Le Territoire du Vide*. Paris Aubier (this edn. translated as *The Lure of the Sea*), by Phelps, J., London 1994, Polity Press.

Foreman, C., 1997. *Fleetwood - Simply the Best*. Fleetwood, Steve Presnail.

Granville, A. B., 1841. *Spas of England and Principal Sea-Bathing Places*. this edn. 1971, Bath, Adams & Dart

Heywood, J., 1916. *Illustrated Guide: St. Annes & Lytham*. Manchester, Heywood.

Illustrated London News, 1856. *The Watering Places of England: Lytham and Southport*. In same, October 4th.

Jealous, A. W., *et al*, 1997. *Lytham St. Annes & the Fylde Countryside*. Lytham St. Annes, GCL, for Tourist Information Centre.

Lytham St. Annes, 1938. *Official Guide*. St. Annes, Fylde Press.

Lytham St. Annes, 1954 & 1959. *Official Guide*. Lytham St. Annes, Express.

Morrison, R., 1997, The nation has need of its hereditary piers, *The Times*, July 12th, 21.

Newspaper Cuttings St Annes Carnegie Public Library. Gazette & News (14.09.1894), City News (08.06.1901), Preston Guardian (11.04.1905), Lytham St. Annes Express (14.01.1927) & (01.07.1927), J. H. Taylor (unknown but post 1920).

Ramsbottom, M., 1995. *Looking around Lytham*. Rochdale, Hedgehog Pubns.

Reynolds, J., (undated). *History of Fairlawn*. Preston, Cranden Press.

Rothwell, C., 1978. *Lytham St. Annes in Times Past.* Chorley, Countryside Publications.

Singleton, S., 1996. *Images of Lytham St. Annes.* Derby, Breedon Books.

Tredre, R., 1997. Oh, I don't like to eat beside the seaside, *The Observer*. June 15th, 16.

Walton, J. K., 1992. *Wonderlands by the Waves - A History of the Seaside Resorts of Lancashire.* Preston, Lancashire County Books.

Wickers, D., 1997. Beside the Seaside - The Solway to the Wirral, *The Sunday Times*, June 19th, 8 – 9.

13 Coastal zone management in the Baltic area

W. H. BALEKJIAN

Introduction

The management of coastal zones involves environmental protection, maintenance of the ecological balance, promotion of economic development and care for the quality of life. It is implemented through the control, neutralisation or elimination of negative factors and the promotion of positive factors at a sustainable level.

Such factors may not be specifically rooted in or limited to, space-wise, a given coastal zone in a narrow sense. They may be traceable to roots in the hinterland of the zone or the quality of the seawater. In this respect the Baltic area is, with particular challenges, not an exception with reference to:

- the almost enclosed (inland) nature of the Baltic sea; a relatively narrow passage links it to the North Sea and the North Atlantic;
- major rivers flowing into the Baltic sea after passing through urban and industrial areas; and
- the existence of 11 independent coastal states making it imperative to manage the Baltic coastal regions as a whole on the basis of international co-operation and partnership.

Thus, while having all the general ingredients associated with coastal management (environmental control, erosion control, ecological protection etc.) coastal management in the Baltic area has geographically and politically specific dimensions, which are noteworthy.

After the disappearance of an inland ice cover some 10,000 years ago, a sweet water lake of ice water remained for some 1,000 years before the land barrier separating it from the Atlantic waters was flooded by sea water, in the form of a huge flood wave from the West. After another 1,000 years, the level of land (presently in South Sweden) rose and once more the link with the Atlantic waters was lost. No more salt water could reach the area; instead, a huge affluence of smelting ice water diluted the concentration of salt in what would become the Baltic Sea. A second sweet water lake was formed for a further 1,000 years. The (currently existing and only link with sea water in the West) Öresund passage became gradually shallower and narrower, generating specific flora and fauna survival problems at the meeting line between sweet and salt waters. Today the Baltic is an ecologically and environmentally sensitive sea and the protection of surviving flora and fauna (including sea eagles, seals, otter, perch, eels, *inter alia*) requires particular knowledge and attention.

The not quite enclosed, not quite inland Baltic Sea is, in the north, surrounded by high hilly coasts and in the south, by rather flat coastal areas. It has a number of large islands like Ruegen (Germany), Bornholm (Denmark), Celand and Gotland (Sweden). It is studded with a profusion of many inhabited as well as uninhabited smaller islands, prominently in the archipelagos of the Stockholm region and the Alan islands between Sweden and Finland. They are all part of the Baltic coastal management scene.

With a water area of 415,125 square kilometres and a water volume of 21,714,000 million cubic metres it needs 25-35 years for a complete renewal of its waters. In the early 1990s it was assessed that one-quarter of its seabed was no longer able to support life, while other levels of the waters were also threatened. Yields from fishing were markedly diminishing; shellfish was considered to be dangerous and inedible; eels were threatened with extinction, and bathing was ruled out in many parts. However, since the beginning of the 1990s the situation has been gradually improving, not least in the wake of campaigning by non-governmental organisations (NGOs) like Greenpeace and as a result of parallel implementation programmes and actions at governmental levels.

The Baltic Sea, an inland sea with dilute salt and almost sweet water contents, could be also described as a complex fjord, firth or gulf with complex subdivisions, separated by many low seabed thresholds. It is a relatively young sea that has been, since times immemorial, a navigational link between its ethnically and politically different communities; it is now becoming, for the purpose of coastal management, an integrated area within the larger framework of the European Union and European co-operation in general.

In an economic and industrial context as at present, the challenges that coastal management in the Baltic area has to face assume, so to speak, macro-dimensions, because they extend far beyond the immediate geographical dimensions of coastal regions. They comprise an area, which extends up to 400 kilometres southward into Central Europe and equally far eastward unto Russia, beyond the three small Baltic republics. This is due to major rivers like the Vistula (Poland) and Oder (Germany and Poland) serving as drainage channels for industrial and other emissions. The river Vistula has been carrying industrial effluents and polluted waters from some 400 cities, villages and industrial areas. Since the middle of the 1990s control and improvement measures are being implemented, not least with due regard to the intention of Poland to improve her environmental protection standards as a candidate for membership in the European Union.

On the socio-economically, better developed, western side of the Baltic Sea in the Swedish coastal regions, emissions of sulphates and phosphates, the excessive use of fertilisers and motorboats have been detrimentally affecting the quality of the soil and waters. Further into the sea dumped war materials, ammunition and poison gas shells have been, and are, further hazards threatening the health of the seabed as well as of the waters above it.

All in all the drainage area served by the Baltic Sea remains four times larger than the sea itself. Some 80 million people live in the larger area in question covering 1,650,000 square kilometres. It covers even parts of the Czech Republic and Slovakia. A phenomenon, which affects the management of the coastal regions, concerns acid rain. This is brought by dominantly southwesterly winds, from as far as Belgium (2%), France (4%), the Netherlands (5%), the United Kingdom (11%) and areas of the former Soviet Union (9%) (1990 estimates). 1997 recorded a reduction of 15% in phosphate emissions. Certain parts of the beaches on the southeast Baltic coast are still not suitable for bathing. While the goodwill to improve the situation is undeniably there, enormous economic, legislative, political and technical (training and institutionalisation) problems exist, to be solved through all-Baltic co-operation, assistance and partnership.[1]

Fisheries are of significant economic importance for parts of the coastal population. Serious concern for the sustainability of fishing communities' interests led already in 1974, to the establishment of a fisheries commission by the coastal states, to deal with the allocation of fishing quotas and the maintenance of ecological balance between the different species of sea fauna. Concerning the, so to speak, health of the

Baltic Sea, the Helsinki Convention of 1980 (revised April 1992) provided for regular monitoring to control the presence of heavy metals and poisonous substances, of substances originating from the coasts and polluting the water. The application of new technologies was placed on the agenda of action, which included the non-use of harmful substances, co-operation to avert leakages and avoid dumping poisonous substances into the sea. A Baltic Sea Institute has been active in Karlskrona (Sweden) since 1992 for research and co-operation between the contracting partes of the Baltic Sea Convention of 1992. It comprises in addition to Sweden, Denmark, Finland and the three Baltic states Estonia, Latvia and Lithuania, and also includes Poland, Russia, Germany and the European Union (EU). EU co-operation and partnership is introducing, with relevance to coastal management in the Baltic area, an increasingly all-European dimension besides that offered by the co-operation within the framework of the Council of Europe (Strasburg) (see below). Such co-operation has proved to be invaluable, particularly in times of unforeseen catastrophes, as in the case of heavy floods in July 1997 in Poland and Germany, when a wave of polluted waters ended into the Baltic Sea.

Discussions between the coastal states on the environmental and. ecological situation in the Baltic area, including matters related to the management of the coastal areas, are gradually assuming a socio-economic dimension. In May 1996, at the Baltic region summit meeting in Visby, Gotland (Sweden), the participating 11 states included in the meeting's agenda *inter alia* and not least the sensitive and complex topic of ecological equality in the Baltic region. The purpose of this approach is to undertake collective efforts to improve the environmental and ecological situation there where it is more urgently needed, independently from limits imposed by national economic or budgetary limitations. This seems to be a commendable approach. Symbiotic relationships between the different regions of the Baltic exists and cannot but have ultimately positive or negative effects in the whole region as a unit.

Favourable developments in the integrational approach to environmental and ecological management in the Baltic Sea region have been promoted by the extension of the EU into the region and the initiation of EU activities. Denmark, Germany, Finland and Sweden are EU members; Poland and Estonia are on the list of the planned enlargement in the coming decade. Already in 1996 at the May 1996 Baltic summit in Visby, Gotland (Sweden), EU Commission President Jacques Santer presented a programme of support for development through co-operation in

197

the Baltic region. A budget of about £3,000 million has been allocated for it and EU member states have stated that the Baltic Sea area is a region of importance for the whole of the EU. This development will induce the promotion of harmonised environmental laws, policies and strategies for the Baltic coastal states within the framework of the EU legal order, with due regard to the EU principle of subsidiarity.

With respect to coastal management, it is worthwhile to refer to the Union of the Baltic Cities (UBC) with a secretariat in Gdansk (Poland). Any coastal city in the Baltic area or other city interested in the Baltic Sea region and its development, may become a member of the UBC. Initially the activities of the UBC will consist in the exchange of know-how and experience.

Returning to coastal zone management in a European context, the activities of the Council of Europe (already mentioned above) deserve more attention.[2] With more than 40 member states, the Council of Europe includes, in its manifold activities at the level of inter-governmental co-operation, work on spatial management and its environmental impact within Europe as a whole. Particular attention has been and is being paid to challenges and problems specific to central and east European countries, some of which are in the Baltic area. They will benefit from additional support by the Council of Europe, even if the Council focuses not least on problems and needs in the Mediterranean area. A Council Action Plan 1996-2000, identifying 11 action themes, includes specific reference to coastal and marine ecosystems (Theme 5) and the establishment of a Pan-European Ecological Network. Action Theme 5 (coastal and marine ecosystems) includes the development and implementation of a European coastal and marine ecological network. (It should be added that the Council of Europe and the EU co-operate closely.) The Council's Committee of Ministers[3] has adopted a number of significant texts on coastal areas and maritime regions.

Coastal regions are associated with human activities involving recreation, holidays, old-age retirement, archaeological preservation, agriculture, natural parks, dumping-grounds etc. Some of these activities have a beneficial impact on the environment and ecological balance; others have obviously harmful and undesirable consequences. With such possibilities in mind, in designated coastal areas in Sweden no new industries are to be established or holiday houses built in particularly sensitive and vulnerable coastal areas. No new constructions may be approved up to 100 metres from the coastline, and this prohibition may apply up to a distance of 300 metres. Exceptions are possible and may be even imperative in some areas, for example, in the Stockholm archipelago zone with its labyrinthian distribution of numerous small islands and the desire of many to secure a secluded small corner on one of them not far from the waterline. In Finland protected coastal areas may be dedicated for use as national parks or fauna havens.

Coastal management in the Baltic area is closely linked with the management of the Baltic Sea as a whole and of the related hinterland areas as a whole, owing to the relatively small geographical extension of the sea, not least as an almost enclosed sea. Thus, coastal management in the Baltic area is part of the larger topic of regional (land + coast + sea) management. This involves, not least in the Baltic area, inter-governmental or international co-operation for satisfactory and sustainable results. In this respect, progress has been satisfactory, even if it is estimated that at the current rate of improvement(s) the better situation, which prevailed in 1950 will be achieved by 2050. There are also new threats to be neutralised. For example, levels of progressive motorisation in Central and Eastern Europe will have to be monitored carefully for checking its effects on air pollution also in coastal areas. This seems feasible in the light of new pollution control and environmental protection technologies. Such resources have been successful in checking or slowing down erosions. Within the sea itself, cleaner water quality has limited, as reported in August 1997, the proliferation of algae. A project to lay a seabed high-tension cable for linking electricity supply systems between the northern and southern coasts of the Baltic Sea is under scrutiny for its compatibility with ecology. It is purely a commercial proposition, with the capacity for 600 MW as part of the Baltic (energy exchange) Ring.

Coastal zone management in the Baltic area has been rightly understood as a matter of regional partnership and co-operation. This approach has halted the deterioration of the situation, which prevailed between the 1960s and 1990s, and it has generated positive results in protecting the quality of the coastal zones (soil, air and water) and progressively improving it. Much remains to be done, dependent, however, not only on political will but not least on the availability of human as well as other resources, the generation of which will be a challenge for decades to come. Thereafter, the situation in the Baltic region will be an integral part of the European regime of environmental and ecological quality endowed with sustainability.

Notes

1. On the state of coastal waters in the Baltic area, see Baltic Sea Environmental Proceedings: First Assessment of the State of the Coastal Waters of the Baltic Sea (1993).

2. Information referred to in the present paper is based on the contents of a paper by Ms. Maguelonne Déjeant-Pons for the 26th International Symposium of the European Faculty for Land Use and Planning (Strasbourg), 15-17 September 1997 in Portsmouth. Ms Déjeant-Pons is from the Directorate of Environment and Local Authorities of the Council of Europe (Strasbourg). This is published as Chapter 2 in this publication.

3. For example, see Resolution (73) 29 (October 26, 1975); Recommendation No. R (84) 2 (January 25, 1984); Recommendation No. R (30) (October 12, 1990). In May 1997, Michel Prieur, from the Interdisciplinary Research Centre for Environmental, Land Use and Urban Planning (France) submitted a draft model on coastal protection within the framework of the Council of Europe (PE-S-CO (97) 2). See also Council of Europe: Report of the European Seminar on the Development and Planning of Coastal Regions. Cuxhaven (Germany), 7-9 May 1985. Study Series European Regional Planning, No. 48 (1986); Council of Europe: Naturopa Special Issue on Coastal Zones in Europe, No. 67 (1991).

14 Database to estimate changes of land use along the Hellenic coast

J. KIOUSSOPOULOS

Introduction

The international concern for coastal conservation is continuously rising, especially after the extensive involvement of the sustainable development principle in the environmental policy. Simultaneously, the idea of multi-sectoral approach of coasts seems to give the necessary umbrella for wisdom and effective actions. On the other hand, it is widely accepted that the very long Hellenic coastline needs a regulation of the conflicting land uses along it, as the previous general efforts in the field of spatial policy have no visible results.

The aim of this paper is to discuss some ideas about the structure of a database aiming at the monitoring and the estimation of land use changes in the Hellenic coastal area. The attempts have been concentrated rather in tracking out an appropriate framework, which could identify the most represented variables and the most useful data sources, than in a detailed description of the database. As land uses are the outcome of competition of many factors, this argument will provide at least a starting point for the organisation of the needed interdisciplinary and other linkages. Finally, it is expected this database not only to be an inventory of the coastal situation, but also to offer a substantial and continuous assistance to the decision-makers concerning the littoral.

The Hellenic Coast

Hellas is the south end of the Balkan Peninsula. The area of the country is about 130,000 sq km, and the total population 10.5 million. More than two thirds of the country is mountainous. The exact percentage distribution of the area according to mean altitude and population is illustrated in Table 14.1.

Table 14.1 The population distribution of Greece

	Area	**Population**
Mountains	42.3%	9.2%
Semi-mountainous	29.0%	21.8%
Plains	28.7%	69.0%

Source: National Statistical Service of Greece

The general distribution of land according to data from the agriculture/livestock census of the year 1991 is illustrated in Table 14.2.

Table 14.2 Distribution of land in Greece

Category of land	**Percentage within Greece**
Cultivated	29.9
Pasture	39.5
Forests	22.3
Inland waters	2.3
Built-up areas	4.0
Others	2.0

Source: National Statistical Service of Greece

The state is divided administratively into 13 NUTS II regions (Ministére de l' Intérieur, 1994) according to the nomenclature of the European Union, four of them consisting exclusively of islands. Only one region does not have a coastline. This is further sub-divided into 51 prefectures (NUTS III – EU nomenclature) of which only eleven do not have a coastline. Within the prefectures there is a local level consisting of 6,000 municipalities or communities. The division of the country in terms of the population of inhabitants is illustrated in Table 14.3.

Approximately a third of the population of Greece live less than 2 km from the coast and as many as 85% of Greeks live within a 50 km radius of the sea (Ministry of Environment Physical Planning and Public Work, 1996). Along the coastline of Greece, there are a remarkable number of islands, which actually constitute approximately one fifth of the total land area (18.9%) and upon which live some 1.5 million people (14.5% of the total population of Greece).

Table 14.3 Distribution of the population of Greece in Municipalities and Communities

Population bands	Municipalities (Total population)	Communities (Total population)
< 100,000	8 (1,972,843)	0 (0)
10,000 – 99,999	119 (3,865,5722)	1 (10,275)
5,000 – 9,999	60 (448,682)	17 (104,995)
2,000 – 4,999	102 (399,149	173 (510,612)
1,000 – 1,999	35 (73,652)	496 (751.436)
500 – 999	27 (39,588)	1,180 (964,685)
200 – 499	8 (5,984)	2,005 (824,838)
0 – 199	2 (1,056)	1,688 (286,383)
Total	**361 (6,806,676)**	**5,560 (3,453,224)**

Source: National Statistical Service of Greece

In addition, the whole state has a very long coastline that is about 15,000 km in length, of which more than 8,000 km consists of the islands. It is the longest coastline of any Mediterranean state (UNEP/MAP, 1996). The coast attracts the majority of the population, and the coastal density is close to 90 persons per sq. km., while the density in the whole country is about 80 persons per sq. km. Moreover, it is expected that the coastal areas - especially the tourist resorts

> are likely to experience significant population increases in parallel with wide fluctuations in numbers of residents from season to season. (Ministry of Environment Physical Planning and Public Work, 1996).

According to the criterion of predominant land cover, there are three main types of coasts. The first consists of about 1,000 km of beaches and sand dunes, where the majority of human activities are located. Almost the three-quarters of the coastline is covered with rocks, while there is a great variety of wetlands (deltas, lagoons, estuaries, etc.) (Museum Goulandri of Natural History, 1994). The Hellenic coast is regarded as national wealth because of the existence of plentiful natural fauna, flora, coastal landscape, sunny weather, clean waters, etc. and manmade resources such as ancient monuments, the attractive built environment, tourist infrastructure, etc. (OECD, 1983).

On the other hand, three categories of economic activities along the Hellenic coastal areas are recognised: built up areas with or without the presence of second sector, tourist areas and agricultural or natural areas. According to data of early 1980s

> over 57% of the total population, 35% of agriculture, 90% of tourism and more than 80% of the industrial activities of the country are concentrated within the coastal zone. (Ministry of Co-ordination, 1981).

For the tourist development in particular, it is indisputable that the attractions of the coast are the main reason the great majority of the 10 million tourists, who visit Greece every year, prefers the coastal areas. As a consequence of that fact, coasts are considered to be of high economic importance, mainly in the smaller islands where the development depends excessively upon tourism. Hence, the establishment of a sustainable development process in their coasts is of great importance for the local economies (UNEP/MAP, 1996).

In parallel to the international tourism, there is a desire of Greeks to own a second home near the sea, and this combined with the international tourism has resulted in a gradual urbanisation of the Hellenic coast. This double pressure produces a strong transformation process and, as a result, forests and agricultural land near the coastal urban centres are being replaced by tourist infrastructure and facilities or becoming resort areas and second home villages (CEC, 1991). This 'land consumption' is the consequence of factors such as economical prosperity and technological advance, which act to influence the land use changes in Hellas (Grenon & Batisse, 1989). On the other hand, conflicts are aroused among other coastal land uses and, according to the opinion of the more pessimists, the final result seems to be evident: the deterioration of the ecological, economic and social value of the Hellenic coastal areas.

The coastal management in Greece

In spite of the long coastline and the maritime character of the Greeks, there is no specific legislation or integrated policy for the Hellenic coastline (Giannakourou *et al.*, 1993; MEPPPW, 1996; MEPPPW - University of Thessaly, 1994; Nikolaou, 1995; OECD, 1983). Despite of that lack, many (usually sectoral) efforts have undertaken in regard to coastal zone policies. A related chronological list for the last two decades is:

1975.   The state's obligation for protection of natural and cultural environment was included in the new Constitution.

1976.   The Hellenic Parliament voted a Law (960) for 'physical planning and the environment'. To this date, planning of this nature was only being controlled from the directions given in the five-year plans and the location of the large investments.

1978.   The Ministry for Physical Planning, Urban Development and Environment ('YXOP') was founded, in an attempt to separate the economic issues from the spatial planning issues.

1979.   A national programme for coastal management was established. It lasted 2-3 years and the elaborated programme was published in the official newspaper of the State (1981) as a Decision (No 9) concerning 'guidelines and actions needed for the management of the coasts'. In fact it was a rhetoric setting without any obligation for a related policy.

1983.   The Law 1337 was voted in. It regulates the current planning of urban areas and other aspects of spatial planning in Greece.

1986.   The Law 1650 'for the protection of the environment' was voted in. It is another rhetoric framework demonstrating rather the goodwill than the decision to carry out a severe policy for the environment.

In the late 1980s a number of special spatial studies ('EXM') were carried out. Many of them are related to islands and coastal areas.

1990.   Joint Ministerial Decision for the regulation of issues related to the environmental impact assessments (EIA) of the large constructive projects.

During the early 1990s, a second series of 'EXM' were set up. All of them studied coastal areas. Unfortunately, as in the previous 'EXM', none were ever enacted.

1996.   The Ministry of Environment, Physical Planning and Public Works ('YPEXODE', successor of 'YXOP') in collaboration with the UNEP/MAP organised a workshop about the coastal areas, in Santorini. That action confirms the recently emerged government interest in coastal management.

1997.   A new law referring to the definition of the shoreline is discussed (Ministry of Finance), in a way to assure the public character of the coast and the ownership of the state. This is another indication of the government's interest in coastal areas.

Furthermore, the above basic laws (360/76 and 1337/83) are currently under amelioration review (Giannakourou *et al.,* 1993; Voulgaris, 1993). On the other hand, under the present conditions, the existent legislative tools, which can be applied along the Hellenic coast, are as follows (all provided from the Law 1337/83):

- administrative enlargement of the coast (up to 50 meters) on behalf of the public;
- prohibition of fences in a zone of 500 meters from the sea, in a way to assure the public access to the coast;
- establishment, wherever it is needed, of a zone of controlled the development of settlements ('ZOE'). It can cover a considerable area and aims to anticipate future development by building and land use control. It is the more integrated among the existent tools, but its implementation is not considered as crowned with success (MEPPPW – University of Aegean, 1995).

In parallel, the following facts are very essential constraints in the implementation of an Integrated Coastal Areas Management (ICAM) along the Hellenic coast:

- absence of a national cadastral information system, which registers all the (extremely small in Greece) properties. Only during the last two years have some pilot projects (Law 2308/95). The whole inventory is planed to carried out by 2009;
- there is neither comprehensive and integrated physical planning nor a national land use plan that cover the entire country (MEPPPW – University of Thessaly, 1994). Instead, *ad hoc* decisions have been made in the case of the location of an industry, a hotel, etc.;
- however, it is legal to build outside the designed urban areas, only if someone owns a big enough property (Wassenhofen, 1995). In most of the cases the critical limit is 0.4 ha.;
- the involvement of numerous Ministries in the coastal issues, each of which applies its own policy. Apart from the already mentioned Ministry of Environment, Planning and Public Works, indicative cases are the Ministries of National Economy, Agriculture, Civilization, Commercial Marine, National Defence, etc. (Spanou, 1994). This multi-sectoral involvement is perhaps the reason that the almost huge number of studies for the Hellenic coast (since the early 1960s) has never been enacted.

All the problems identified above have cultivated among the citizens the conviction that it is permitted to build everywhere - even if they have not the legal right (Getimis, 1989). The coastal areas are the most preferred domains for the realisation of the previous belief and the politicians usually encourage this tendency, since they do not take strict and effective measures against it (MEPPPW – University of Thessaly, 1994). Besides the previous phenomenon/inclination, the big projects and especially the transportation networks - which are currently under construction - will change the spatial image of the Hellenic territory (Loukakis, 1993). This fact produces new development potential and, as a consequence, the coastal areas are likely to be even more stressed. It is worth noting that every summer, fires destroy large areas (200 – 800 sq. km.), many of which are coastal forests. Since the land regulation is not well coordinated, many of the burnt areas eventually become built upon and urbanised, although there is a relative constitutional problem.

Database: methodological/structural problems

After adopting the necessity of an integrated approach of a coastal area and identifying the goals and the objectives of such a process, the first preparatory action is the exact *definition of the studying area (A)*. In addition, the capital of Hellas was just elected as the city that would organise the Olympic Games of 2004. This will probably create extra strain on the Hellenic coast. It is not a simple problem, as it can obtain more than one resolution (CEC, 1991; Kioussopoulos, *et al.,*1997; OECD, 1993). In any case, it is proposed that two levels of approaching the definition should be adopted. The first is confined in a narrow ribbon of land, where the pressure of development process is intensive and the risks of environmental degradation are obvious and undoubted. Apart from the previous critical zone, the second level of coastal approach is related to a wider area, which includes terrestrial and marine sections. A more detailed procedure can identify a third even broader area of direct or indirect influence to the other two zones (UNEP/MAP, 1996). The previous decision of the coastal area width is connected to the (technical) theme of the selection the *smallest appropriate area unit.*

After defining the area, the next step is to choose the *variables (V)* to be observed. Since the central objective of the proposed database is land uses, a variety of uses must be the first group of parameters to be studied. Linked to the land uses are the (natural and manmade) resources, which attract the human activities. In addition, infrastructure and pollution inventories seem also to be meaningful for the database. Connected to the previous approaches are the (technical) issues of selecting the suitable sources of data, kind of database management, map scale, etc.

It is self-evident that the above-described fundamental components of the database create an instant representation of a coastal area. If the acquisition of another time picture is desired, the database should be fed with the appropriate data. Also, a (technical) issue for the whole database is to define the *time intervals (T)* between the variable observations. Moreover, if it is preferred, the database could supply knowledge about the past figure of the examined coastal area, then it should be decided how long in the past the coastal area should be studied.

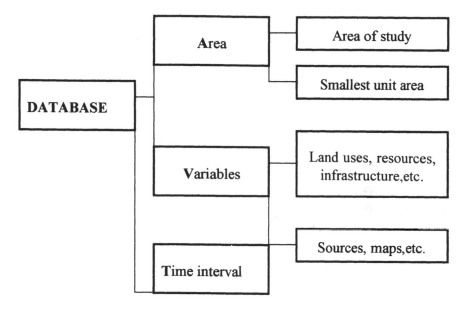

Figure 14.1 The database at a glance

Database: some thoughts/proposals

Area

The territory of the Hellenic municipalities or communities ('OTA') can be applied as smallest unit area. The proposed smallest unit area refers completely to the terrestrial part of a coastal area, but it does not mean that the maritime part is absent. As the human beings live on the lithosphere their economic activities all around the earth reflect ultimately to the land uses on the lithosphere. In any case the proposed unit area could be extended seawards.

Their average area is about 20 sq. km., which is equivalent to a 4.5 km square. The number of inhabitants per municipality or community varies a lot, as it was already shown in Table 14.3. This choice has disadvantages but it can be accepted since it also gives a great flexibility. In fact, it offers the opportunity to investigate different spatial dimensions (widths) of coastal areas, according to a proposed variable, the *proximity* or vicinity (p), defined as following:

- the 'OTA' being immediately adjacent to the sea corresponds to the more critical zone of the coastal area and they have value of proximity equal to 1 ($p = 1$);
- the 'OTA' not having common boundaries with the sea but being immediately adjacent to the 'OTA' with $p = 1$, participate to a wider coastal area and they have $p = 2$;
- the 'OTA' which do not have common boundaries neither with the sea nor the 'OTA' with $p = 1$, but are immediately adjacent to the 'OTA' with $p = 2$, compose a more broad coastal area ($p = 3$).

In addition to the previous possibility, it is very important to acquire the existent of the availability of data for the 'OTA', as there is not yet a cadastral survey in Greece. If data is available, it is possible to apply all the benefits applying GIS process (Kioussopoulos, 1997). On the other hand, a unit area with more normal boundaries, e.g. a 500 meters square (EC, 1993) does not necessarily carry information as it has an imprecise location. Thus, it is serviceable for only what is 'watched' (e.g. in a satellite images), without the possibility of 'perceiving the background' (human activities, social competitions, etc.). Finally, the 'OTA' is not a particularly good choice as the smallest unit area (see the disadvantages in Table 14.4), but it seems to be an adequate and helpful selection.

Variables

The most important human and economic activities along the coastline are agriculture, forestry, fisheries and aquaculture, industry, energy production, tourism and recreation, transportation and settlements (UNEP/MAP, 1994; UNEP/MAP, 1996). All of them consume coastal area resources and above all they need areas of land. Moreover, they need infrastructure and they product pollution. The more common way to evaluate the extent of the previous phenomena and variables is to measure the occupied *area*. As this approach is very shallow and only seldom represents the real potential of each land use, it is suggested that more penetrating parameters, such as the following be utilised:

- measuring and assessing the number of inhabitants, the people working in each category of land use, the invested capital, the value of the land, the production of each activity, the number of hotel beds, the available quantity of every kind of recourses, the kind of transportation network, the kind of emitted pollutants and many other similar and easy to determine variables;
- formatting indexes from the previous variables. Moreover, as the absolute values of data are usually without real meaning or with poor qualitative information, it is recommended relative values of variables be utilised.

Table 14.4 Relative merits for using 'OTA' as the smallest unit

'OTA' (Municipalities and Communities) as smallest units	
Advantage	**Disadvantage**
Availability of data Alternative methods of delimiting boundaries Existence on digital maps Already used for other purposes as a 'smallest area' Co-operation at a local level	Not equal in area Random spatial distribution Boundaries may not appear on aerial photography Boundaries are not likely to appear on satellite images Boundaries do not appear on all maps Only represent the terrestrial area and no maritime part

The number of potential variables listed above proves the need of identifying the probably existent linkages among them. Furthermore, a critical point is the query of methods to *estimate accurately the non-quantitative phenomena, which are related to the coastal land uses*. In fact, as the coastal areas are characterised by complex patterns of interaction between natural and human ecosystems, and its land uses are the outcome of the competition among the potential users, *more complicated functions should be involved in the database*. Some of them could be the number of benefited people, the social class or the economic power of them, the size and the multi-nationality of the involved enterprises, the other existent job opportunities, the recoverability of the utilised natural resources, etc. As a consequence, it seems to be necessary the establishment of a *human-interface* between the real world of the coastal area and its appearance at the database.

In any case, the list of variables can never be exhaustive, although there are methods to classify them (UN, 1991). In parallel, it is common sense to reduce its number by customising it to the studied coastal area. Finally, each land use can be specified as deeply as it is desirable, according to its importance in the observed coastal area, e.g. tourism in a coastal destination (CEC, 1994; MEPPPWork – University of Aegean, 1995).

A special concern should perhaps be directed towards *the means of transportation*, as the next question to emerge: are new road infrastructures (routes, highways, ports, airports, parking facilities, etc.) the predominant motive factor of pressure to the Hellenic coast?

Of equal importance is the number of observed variables, as complex is the variety of *data sources*. An overview of that available in Greek sources (in connection with 5 properties) is illustrated in Table 14.5.

In particular, the use of aerial photographs to track the new 'roads' is an easy and relatively cheap way to scan the changes in local level (with the help of existent maps). There are plenty of them for the Hellenic coast, usually in scales greater than 1:10.000. In addition, they are reliable inventories for a precise point of time and offer many categories of data (observed in foreground) without the need of trained staff. Furthermore, local authorities could also use the database.

Time

As the time period between censuses in Greece is ten years, a decade is the suitable (larger) time interval for the updating of the database. In addition, an intermediate five-year period can be applied. It is obvious that in order for the database to be deeply useful it must be regularly updated. This can take place continuously, but a report as often as is needed (e.g. 2-3 years) seems to be a realistic goal.

Conclusion

The adoption of 'OTA' as the smallest unit area is a quite apt and practical choice, despite its disadvantages. In the field of variables, it is of great importance to create compound indexes in a way to focus on the 'background' of coastal land uses and to emphasise to the real motivators who convert them.

The utilisation of aerial photos could be very functional and effective. It supplies reliable knowledge about the *past* image of the examined coastal area, reduces the uncertainly about the *present* situation and could *foresee* the expected impacts along the Hellenic coast, especially those which arise from the new roads projects.

It is obvious that further research should be undertaken firstly to investigate deeper the effectiveness of the proposed combination 'OTA-aerial photos' in the Hellenic coast and secondly to explore the human and economic activities in a way to find out more represented variables connected to the coastal land uses.

References

CEC, 1991. *Europe 2000 – Outlook for the Development of the Community's Territory*. Brussels - Luxembourg.

CEC, 1994. *Taking Account of Environment in Tourism Development*, Study, Econstat, Luxembourg.

EC, 1993. *CORINE Land Cover - Technical Guide*, Luxembourg.

Getimis, P., 1989. *Policy about Human Settlements in Hellas – The Limits of Reformation*, Odysseas, Athens.

Giannakourou, G., Gartzos, K., Bazou, F., Beriatos, H., & Papaioannou, G., 1993. *Collection, Evaluation and Annotation of the Legal Framework about the Physical Planning in Hellas – Proposals and Directions for a new Framework*, Study, Technical Chamber of Greece, Athens.

Grenon, M., & Batisse, M., (eds.) 1989. *Futures for the Mediterranean Basin - The Blue Plan*, Oxford University Press, UK.

Kioussopoulos, J., Economou, Cl., & Gikas, T., 1997. Coastal areas: an approach with GIS. In: the scientific meeting *From Surveying to Geoinformatics*, TEI, Athens.

Loukakis, P., 1993. Tendencies of transformation in the Hellenic spatial structure. In: the international conference *Greece in Europe – Spatial Planning and*

Regional Policy towards the year 2000, Technical Chamber of Greece, Athens.

Ministére de l' Intérieur, 1994. *Division Administrative de la Grèce*, Athènes.

Ministry of Coordination, 1981. *National Coastal Management Program – Greece*, Athens.

Ministry of Environment Physical Planning and Public Work, 1996. National Report of Greece. In: the *Workshop on Policies for Sustainable Development of Mediterranean Coastal Areas*, Santorini.

Ministry of Environment Physical Planning and Public Work – University of Aegean, 1995. *Coastal Management – Consequences of Tourism*, research project, Athens.

Ministry of Environment Physical Planning and Public Work – University of Thessaly, 1994. *Outline and Means of Control for the Land Uses in the Non-Urban Areas – The Situation in other States, Proposals for Hellas*, research project, Athens.

Museum Goulandri of Natural History – Greek Biotope/Wetland Center, 1994. *Inventory of Greek Wetlands as National Resources*, Thessaloniki.

National Statistical Service of Greece, 1991. *Distribution of the Country's Area by Basic Categories of Land Use (1991)*, Athens.

National Statistical Service of Greece, 1996. *Statistical Yearbook of Greece 1994 - 1995*, Athens.

Nikolaou, A., 1995. *Environmental Aspects in the Planning of Coastal Areas*, mimeo.

OECD, 1983. *Politiques de l' Environnement en Grèce*, Paris.

OECD, 1993. *Coastal Zone Management – Integrated Policies*, Paris.

Spanou, K., 1994. Administrative networks for environmental policy. In: *Environment and Law*, Vol.1, Athens.

United Nations, 1991. *Concepts and Methods of Environment Statistics, Statistics of the Natural Environment.* Technical Report (F/57), New York.

UNEP/MAP, 1994. *Guidelines for Integrated Management of Coastal and Marine Areas (with special reference to the Mediterranean basin)*, Split.

UNEP/MAP, 1996. *The State of the Marine and Coastal Environment in the Mediterranean Region.* Technical Report (No 100), Athens.

UNEP/MAP, 1996. *Workshop on Policies for Sustainable Development of Mediterranean Coastal Areas.* Technical Report (No 114), Athens.

Voulgaris, A., 1993. Physical planning: efforts-problems-perspectives. In: The International Conference, *Greece in Europe – Spatial Planning and Regional Policy towards the year 2000*, Technical Chamber of Greece, Athens.

Wassenhofen, L., 1995. Physical planning and the country. In: The Scientific Conference, *Regional Development, Physical Planning and Environment in the Framework of the United Europe*, Hellene Regionalists Association, Athens.

15 The problems of coastal zone management in Turkey

F. NARLI

Introduction

Coastal Zone Management (CZM) can be defined as establishing policies for the most realistic and long-term usage of the resources of coastal areas which show geographical diversity with sea and land interaction and have natural, cultural, ecological and economic significance. Moreover, it aims to determine the principals of planning and programming and their application techniques in coastal areas with high participation and integrated effort. This sort of a management necessitates not only a well established institutional organisation where the authorities, duties and responsibilities of the participants have been clearly defined but also necessitates a consistent, sufficient, applicable law, democratic and realistic planning and an established application mechanism. Coastal Zone Management in Turkey will be examined in this context.

A non-structurised institutional organisation

In Turkey, planning, application and monitoring of the plans and furnishing of the public services have not been handled in an integrated management strategy. In this respect, there is not a national organisation comprised of the authorities of the responsible institutions, that is developing management politics in national scale, addressing the demands, putting the targets, programming the priority actions. Similarly, there are not any regional and local organisations under which the authorities responsible of coastal zone management have been presented, and defining the problems and needs of these areas, determining the priority regions and the settlements, preparing the priority action plans, regional development and management programmes.

Therefore, many different organisations are responsible for various steps of the management process. These include the Governmental Planning Authority and some ministries, together with their general directorates. Also, general directorates associated with public services may participate the management without any organised collaboration among them and without having a well-defined national or regional management policy and programme. For example, the Ministry of Tourism is responsible for the planning of 'Developing Areas for Tourism' and from the approval of those plans. Similarly, the Ministry of Environment is responsible for those related to 'Specially Protected Areas', the Ministry of Forestry is responsible of those related to 'National Parks' and the Ministry of Culture is responsible of those related to 'Areas having Natural and Cultural Wealth'. Each of these organisations is responsible for the preparation and application of plans for the related region but do not have any communication between while doing this.

An inefficient communication exists among the ministries, between the ministries and the authorities responsible for providing public services and even in the self-organisations of the authorities. For example, ministries do not usually act jointly with public offices while they are preparing and approving the plans for their areas of interest. However, public offices should supply technical and social equipment to the same areas even if they do not participating the planning phase. Similarly, there is not an efficient collaboration among those public offices responsible for the planning, application and running of the services. As a result, many activities can be repeatedly organised and thus reduce the effect of each other. As an example, while the problems of land can be dealt individually by Directorate of Forestry, Directorate of Agriculture, Directorate for Rural Services, there is no authority for the protection of beaches. Additionally, the projects of drainage, drainage of the wetlands, construction of highways and dams in an area have not been handled in regional scale with a management perspective but instead left to the decisions of public authorities. Here, it has to be emphasised that all the authorities responsible from the management of the coastal areas have their own budgets, labour and technical equipment. Each authority provides their own policy for the planning, application and serve while trying to protect their own organisational principals. Several organisations might be asked for duty for the same problem and this causes chaotic situations in bureaucracy and prolongs the period. On the other side, the unsystematic way of management applications by different authorities prevents one single authority from pursuing a problem from beginning to end. Some of the authorities might also be disregarded as a consequence of unsystematic applications. As an example, there is not such an organisation to conduct research prior to the preparation of plans and projects, observing the difficulties in application phase and making assessments about the impacts of the projects. All these prove that the central governmental authority plays the main role in coastal zone management and such an authority creates disorder and inefficiency at different stages. However, the lack of communication and confusion in responsibilities is not only happening in the central authority but it is going on between the central and local authorities, too. Public authorities, which have very restricted role in coastal zone management, have to prepare an application plan for the projects and plans that they have not participate their construction from the first stage. Then they have to apply these application plans and provide all the public services in their municipality zone. There is not any correspondence between the local authorities related to Coastal Zone Management. Moreover, non-governmental organisations, chambers of professions, local people and other consumers, universities and research institutes have not been participating the coastal zone management system.

The above mentioned points oblige to form a well framed administrative organisation which hosts all the authorities dealing with coastal zone management, clearly defines their duties and jurisdiction and looks for a complementary perspective.

A non-defined organisation of law

The law and regulations related to CZM have been examined through their principals, content and sufficiency and the difficulties in application and the results achieved from the applications have been evaluated. Although this should be an inter-disciplinary effort, a general assessment has been given below.

Many organisations authorised for preparing the draft laws have been working independently and may suggest inconsistent drafts as they have a loose communication among them. This creates a big problem in the organisation of law. For example, *The Law of Coastal Protection* has been prepared completely by the Ministry of Reconstruction from their own perspective. In the same sense, the Ministry of Environment and the Ministry of Tourism have prepared *The Law of Environment* and *The Law of Tourism*, respectively. These laws cannot completely satisfy the demands of integrated coastal zone management issue. Sometimes very similar laws or regulations related to the same subject can be assembled and this makes the application and control stages very difficult. For instance, while the Ministry of Environment suggests that the 'who pollutes pays' principal has the support of consumers to the infrastructural expenses, the Ministry of Tourism offers discount for the infrastructure services to the tourism investors in the frame of *The Law of Encouragement for Tourism*. It is also possible to witness the same problem in different items of a law.

The objectives of the laws are not clear enough. For example, the *Law for Coastal Areas* is not matching to urbanised coastal areas. Regulations having some 'exceptions' weaken the laws' clarity and applicability as they are open to argument. The inefficient definitions of some concepts create problems in the application phase of laws. The definition of 'shoreline' is a good example for this. Borders of the shoreline has been defined without noticing the type of the shore and the geographical conditions (e.g. marshes, rocky coasts, sand dunes) and this creates problems while determining the borders of the shoreline for private settling. Similarly, the expressions of 'separated for public usage' and 'economical activities that have to take place in the vicinity of shore' have not been based on geographical, technical definitions and terms of law. *The Law for Coastal Areas* proposes the protection of coasts but it does not comprise an ecological definition, sustainable usage of coastal zone and does not bring any protection criteria. The obligation of preparing Environmental Impact Assessment (EIA) reports which in fact, not so successful in practical applications is not sufficient in protection. The measures taken for coastal area usage have been usually related with terrestrial environment not with the marine environment. While various ministries are declaring specially protected areas in their own attitude, the preparation of a management plan including the ecological budget, the restoration and renovation of the area, daily maintenance, opening the area to public visit have not been provided by law.

A further point is the ineffective application of what is written in the law. For example, the *Law of Environment* and the *Law for Coastal Areas* foresee the preparation of EIAs for every kind of investment in coastal areas. However, the insufficient data sets, time consuming procedure and limited technical services make it and the formation of EIA working commissions difficult. The Law of Environment has charged the local authorities with the monitoring of the environmental quality and control.

However, the local authorities have neither the legal permission nor have the budget and technical services. The *Law of Reconstruction* charges the municipalities for the control of illegal constructions in the municipality borders but the municipalities sometimes can not endure this because of lack of some technical and financial features and even because of insufficiency of positive willing. The present organisation of law is not sufficient for coastal zone management. The main problems in law can be summarised as:

- the confusion in the mechanism of law;
- the frequent renewal of law and regulations and their inconsistent and disintegrated nature;
- being open to arguments and comments;
- not including the finished operations;
- not completely known by the authorities and unwillingness;
- the delays in applications, ineffective punishments;
- insufficient technical, financial and human resources.

The situation detailed so far extremely necessitates an integrated and clear organisation of law, which is capable of managing both usage and protection of dynamic coastal areas.

The inconsistency of plans

In Turkey, there is no national or regional management plan. The administrative organisation, excluding the management concept and the deficiency of management policy for the country or a whole region, prove the actual case. In present, there are two main sorts of plan related to usage of coastal zones. First is the 'National Development Plans' prepared by Governmental Planning Organisation in which general objectives, with respect to economic sectors, have been defined for the socio-economic development of the country. The second is the physical plans (regional plans, urban plans, application urban plans and specific plans) directing the spatial development prepared by authorities having duty.

There is no hierarchy, consistency and integrity among the planning documents. The plans aiming the socio-economic development and the plans for spatial development have been prepared and approved, respectively, by Governmental Planning Organisation and by the related ministries (e.g. Ministry of Reconstruction, Ministry of Tourism, Ministry of Forestry, Ministry of Environment, Ministry of Culture) and the municipalities without any collaboration. Patara (Ovagelemis) is a typical example for this. The region has been accepted as a 'Tourism Developing Area', 'Specially Protected Area', 'Protection Area for Natural Wealth' (being a spawning area of sea turtles) and 'Protection Area for Cultural Wealth' by different ministries and the plans for the same region have been prepared with different perspectives. Therefore, the Ministry of Reconstruction has not approved any of the plans and Patara could not have a management plan for many years.

The lack of integrity, complement, harmony and collaboration in local scale is another problem in the planning stage. Each of the coastal settlements aim to have the best plan responding to their own needs but they do not satisfy the geographical properties, socio-economic realities and the environmental problems of their region. This has consequently caused the creation of many plans which have different priorities, interests and needs. However, planning in coastal zone management necessitates a picture that has not been limited by administrative issues but takes into consideration the spatial and socio-economic development and planning of many settlements together within geographical integrity. On the other hand, the exclusion of rural areas of with populations of less than 3,000 within urban plans, causes discontinuity in the integrity of coastal usage.

In Turkey, the terrestrial environment has limited planning. It does not include decisions on marine usage and protection. There is not a separate plan that organises the usage and protection of economic activities in marine environments. Presently, coastal planning has been limited with some plans and projects prepared by different authorities for different purposes. Although these plans have been able to be documented after long and detailed studies, they have important missing points in view of coastal zone management.

Both the developing plans and the spatial plans have not been based on a data set, which includes the coastal potential, different types of coastal usage, and their impacts. Developing plans are the documents where only the objectives of different sectors are listed. Therefore, they maintain judgements whereby the objectives of the different sectors can conflict with each other. For instance, the tourism encouragement policy is conflicting with environmental and coastal protection policies. However, the developing plans should present a comprehensive objective beyond the individual politics of all the sectors. Additionally, there are conflicts among the directions of long-term, mid-term and short-term developing plans.

In contrast, there is similar inconsistency and conflict between the physical plans prepared for different purposes in different scales. For example, during the preparation of tourism plans, the tourism potential has been envisaged in every detail but the coastal environment (its nature, formation and development) have not been investigated properly. In any of the plans the carrying capacity of the coastal environment, the impacts of the planned activities on natural and human resources and the impacts of urbanisation on the terrestrial and marine environment (water quality, ecological balance of the coastal waters, underwater archaeological resources) have not been investigated. The sensitive and priority regions have not been determined. The natural risk analyses has been restricted with risks of earthquakes and floods but the consequent damages and the precautions have not been analysed.

Economic development and the population projections are poor. Some judgements have been taken related to the protection of traditional activities, agricultural, natural and cultural privacy areas but precautions have not been provided. One of the major deficiencies of the plans is the undetermined public reaction to the existing way of coastal zone management and their future demands. Generally, the expenses for the application phase and the socio-economic demands have been considered in plans whereas the projects related to the regional self-financing have usually been disregarded. In the later step of the application phase, a model for the expenses for the technical and social services and their management has not been developed. The impacts of the plans to the area out of the planning site have nearly been omitted.

The major problems appearing in the application phase of the plans are:

- renewal of plans with certain revisions;
- obtaining poor infrastructural equipment as a result of illegal constructions;
- inefficient inspection and control;
- lack of effective punishments;
- delays in the application of plans; and
- unavoidable speculation of building, land and illegal building.

There is no organised programming of units and the management provides many forms of services, with the exception of the Southern Antalya Tourism Developing Region and Infrastructure Management Union. The limited equipment, funds, personal, data and knowledge are the main problems in the supply of services.

Recommendations

An institutional organisation dealing with coastal zone management has to take its place in administrative mechanism. All the actors dealing with coastal zone planning, application of plans and supply of services together with the users and the researchers have to take their own roles in this institutional organisation with communication, collaboration and integrity. The so-called organisation has to perform national, regional and local coastal management policies and develop the present planning, programming and application techniques. While doing these, it has cautiously considered the coastal zone within the integrity of marine and terrestrial environments and in a complete management perspective. By this way, the sustainable use of resources will be established where the balance of protection-consumption, rational and long-term benefit can be achieved. This organisation should be equipped with financial means.

An organisation of law has to be formed which includes all the criteria necessary for Coastal Zone Management. For this purpose, the problems arose from the present law and regulations and application of them have to be clearly defined, conflicting and repeating ones have to be eliminated, the lacking ones have to be included and, finally, should be updated due to the needs of Coastal Zone Management.

Index